Technician's Guide
to
Solid-State Electronics

Technician's Guide
to
Solid-State Electronics

Morris Grossman

Parker Publishing Co.   *West Nyack, New York*

Library of Congress Cataloging in Publication Data

Grossman, Morris
    Technician's guide to solid-state electronics.

    1. Semiconductors. I. Title.
TK7871.85.G68   621.3815'2   75-42414
ISBN 0-13-898585-5

Printed in the United States of America

Solid-State Made Easy . . .

This book fills an important need by providing the key factors involved in solid-state electronics, showing simply and clearly how to understand and work with these important devices. Every day brings the announcement of another "new" electronic device—IC, MSI, LSI, LED, CMOS. At least on the surface they appear to be new. However, a practical knowledge of solid-state fundamentals can tie them all together for you. You will understand the meaning and significance when a manufacturer claims improved specifications for his "new" device and will get simple and clear explanations of how the latest devices work and how to keep them working.

To confidently and quickly master the large volume of new solid-state devices, you need only a firm grounding in basics. With an awareness and understanding of the fundamentals, you will quickly grasp this area of vital importance to the electronic technician.

Though magazine and manufacturers' information sometimes covers solid-state fundamentals in simplified form, always limited by space, these publications often serve only to tease the reader—since material cannot be treated completely and in depth.

Even the reason for the use of the term "solid-state" is seldom clearly stated. It means simply that solid-state devices make use of some unique property of matter in the solid state. The other states of matter, such as liquid, gas and a rather new one, plasma, are also used in electronic devices—but such devices are differently categorized. They go under such names as electrolytic, gas-discharge and magnetohydrodynamic devices.

To explain such extraordinary underlying ideas as those in quantum mechanics and the wave theory of matter requires a more detailed approach than you can get from the fragmented treatment in magazines. A simplified method of treating fundamentals is of primary importance, however. This book will help you break the barrier of complex mathematics and obscure language.

An inquiring mind, a technician's normal vocational education and a good grasp of elementary mathematics is all you need to follow the contents of this book. More advanced ideas are explained clearly as you need them. Even the jargon of higher mathematics becomes surprisingly simple when explained in everyday language.

You will not only acquire the basic fundamentals of solid-state technology, but this book will relate them to most of the modern solid-state devices available—and others that will soon be forthcoming. The book provides useful and practical information on device characteristics, design ideas and hints, simple tests to perform, applications of solid-state devices and troubleshooting techniques. Although major attention is paid to devices currently in use, the book explores some areas in the idea and experimental stages. Things are moving so rapidly in this area that you will probably be able to apply this information sooner than you think. The keynote of this book is practicality and clarity, and we believe it will help you feel like an insider, in firm command of the knowledge that is essential to success in working with solid-state electronic equipment.

Morris Grossman

"The society which scorns excellence in plumbing be-cause plumbing is a humble activity and tolerates shoddi-ness in philosophy because it is an exalted activity, will have neither good plumbing nor good philosophy. Neither its pipes nor its theories will hold water."

—John W. Gardner

Electronics is more than just components and wires. Components, like vacuum tubes, become obsolete and even wire has been replaced to a great extent by the printed-circuit board. A technician who has learned only the "plumbing" should not be scorned but pitied, because he too soon will become obsolete. But when he grasps the real fundamentals—the "philosophy"—he becomes obsolescence-proof. Both a good "plumbing" knowledge and a strong philosophy back-ground will assure that the technician's work and word both "hold water."

Many books provide the technician with the plumbing aspects of electronics. But the "exalted" status of the philosophy—solid-state physics, the basis of modern electronics—gives electronics the reputa-tion of being a difficult subject, which many technician-level books do not cover in depth. We believe that this book will fill the gap and provide the electronic technician with the essential philosophy he needs to do a good plumbing job.

Contents

13

Technician's Guide
to
Solid-State Electronics

Chapter 1

Mastering the Fundamentals of Solid-State Electronics

Science cannot be built on contradiction as a fundamental principle. Paradoxes must be faced squarely and removed.

Open almost any book on semiconductors and you immediately run into a conflict with "common" experience. You are introduced very quickly to the term "energy gap." This term is used to describe a forbidden energy range that an electron must pass through, in one jump, as it changes from a valence[1] electron to a "free" state where it is available for contributing to electrical conduction (Figure 1-1).

Forbidden energy levels seem to contradict common experience. Why can't an electron have an energy level within this energy gap?

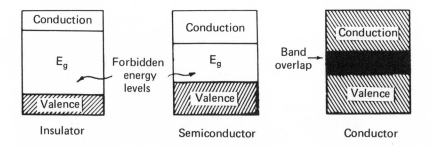

Figure 1-1. Modern wave mechanics explains the difference between an insulator, a conductor and a semiconductor by the size of the forbidden energy gap. Valence electrons must acquire a minimum amount of energy, E_g, before they can contribute to electrical conduction.

---

[1]See pages 31 to 34 for details.

Step on the accelerator and your car can assume any speed from zero to its top limit. There are no energy gaps.

But is our common experience all that one-sided? Does everything we experience possess completely continuous variability?

Let's not ignore the most obvious. Chemists have a chart called the Periodic Table that arranges the chemical elements into families of similar chemical properties. But physically, the basic chemical elements completely lack continuity of properties. Proceed from one element to another along the scale of atomic numbers (Table 1-1). Add an electron to neon (atomic number 10), and you get sodium (atomic number 11). Neon is a gas and sodium is a solid with completely different physical characteristics. There is no continuity of properties between atomic numbers 10 and 11, or between any two other adjacent elements.

Making common sense of the energy gap

Other examples of discontinuity or gaps appear as you dig deeper. Some of them will be discussed later. But for a general introductory explanation, consider the following statements.

- When you add a lot of zeros, you still get zero.
- The whole of any quantity (mass, volume, time, energy, etc.) is the sum of its parts.
- Therefore any quantity cannot be subdivided without limit into zero-sized parts.

The first two statements are logical and common-sense premises. The third statement is a conclusion that must follow. Accepting the conclusion can open your mind to the basic assumption of quantum mechanics—that, in general, the energy of particles can't change in a continuous manner, but the changes are in jumps of a certain minimum quantity. In the complex situation within solids, these jumps become energy gaps of various sizes depending upon the kind of material, its purity and crystal structure, and many other factors, as we shall see.

The energy gap makes the difference

Insulator, semiconductor or conductor of electricity—the energy gap determines the classification. An insulator has a wide energy gap

Atomic number	Symbol	Element	Ionization eV	Number of outer-shell electrons
1	H	Hydrogen	13.60	1 full outer shell
2	He	Helium	24.58	2
3	Li	Lithium	5.39	1*
4	Be	Beryllium	9.32	2
5	B	Boron	8.30	3
6	C	Carbon	11.26	4
7	N	Nitrogen	14.54	5
8	O	Oxygen	13.61	6
9	F	Fluorine	17.42	7†
10	Ne	Neon	21.56	8# full outer shell
11	Na	Sodium	5.14	1*
12	Mg	Magnesium	7.64	2
13	Al	Aluminum	5.98	3
14	Si	Silicon	8.15	4
15	P	Phosphorus	10.55	5
16	S	Sulfur	10.36	6
17	Cl	Chlorine	13.01	7†
18	A	Argon	15.76	8# full outer shell
19	K	Potassium	4.34	1*
20	Ca	Calcium	6.11	2
21	Sc	Scandium	6.56	3
22	Ti	Titanium	6.83	4
23	V	Vanadium	6.74	5
24	Cr	Chromium	6.76	6
25	Mn	Manganese	7.43	7
26	Fe	Iron	7.90	8
27	Co	Cobalt	7.86	9
28	Ni	Nickel	7.63	10
29	Cu	Copper	7.72	11
30	Zn	Zinc	9.39	12
31	Ga	Gallium	6.00	13
32	Ge	Germanium	7.88	14
33	As	Arsenic	9.81	15
34	Se	Selenium	9.75	16
35	Br	Bromine	11.84	17†
36	Kr	Krypton	14.00	18# full outer shell

The left portion of rows 3–10 is labeled "Helium core", rows 11–18 "Neon core", and rows 19–36 "Argon core".

\#Inert gases; do not combine chemically
*Alkaline group; valence is +1
†Halogen group; valence is −1

Table 1-1 Atomic structure—partial table

or forbidden zone so that electrons from the valence band are not easily dislodged into the conduction band. Semiconductors have a smaller forbidden zone so that electrons more easily bridge the gap and can become conductors of current flow. Conductors have very small or overlapping energy levels between valence and conduction states so that very large amounts of conduction electrons are normally present to drift under the influence of an applied potential difference.

Current flow depends upon the number of charge carriers (electrons are not the only charge carriers) and their mobility. The more charge carriers in the conduction bands and the longer (and further) they can move before colliding (called mobility), the greater is the conductivity.

In metallic conductors, applying heat increases the intensity of the random vibrations of the atoms in the crystal lattice. Electrons drifting under the influence of an applied voltage are therefore more likely to collide with atoms and be deflected from the voltage-directed current path. This action is called a reduction of mobility and results in an increase in resistance with increasing temperature, characteristic of metallic conductors. The density of electrons in the conduction bands in metals is so huge to begin with that heat has very little effect upon the conductivity by increasing the number of charge carriers. The temperature coefficient of resistivity (or conductivity) in metals is therefore governed by electron mobility.

In semiconductors, applying heat likewise increases the random vibrations of the lattice atoms. But the semiconductor forbidden-energy gap normally "starves" the material of available charge carriers. Thus a temperature rise produces a significant increase in charge carriers, and the charge carrier increase has a much greater effect than any mobility reduction. The result—semiconductor resistance is reduced with increasing temperature.

In insulators, the energy gap is so wide that the temperature of the usual solid-state device environment can't provide sufficient energy for many electrons to bridge the gap. Thus no conductivity exists.

Explaining the energy gap

The ability of the energy-gap theory to account for conductivity characteristics of insulators, semiconductors and conductors is but another factor supporting the atomistic approach to dealing with energy (or almost any other quantity). If energy can't be divided into zero quantities, then there must be some smallest package. The charge on an electron seems to be the smallest amount of negative charge. All larger amounts of charge are integral multiples. Likewise, energy comes in photon packages. And all larger amounts of energy are integral multiples of photons.

The charge on the electron and the energy content of the photon are so small that in the quantities used in practical (macroscopic) situations continuous divisibility seems to hold. This seeming continuity of things is almost an instinctive viewpoint. The continuum concept was unquestioned by science for a long time. In fact, so great was the aesthetic appeal and so close was the continuum approach to the "common" experience that it was not until well into the twentieth century that scientists were ready to admit that something was wrong.

Scientists, satisfied with their theories, were only looking to improve the accuracy of that next decimal place, but all sorts of things continually seemed to go wrong with the experiments. Mathematicians had learned to deal with the ultimates in the abstract. They had invented the infinity symbol (∞) for the ultimate in bigness and zero for the extremely small. But many scientists soon began to suspect that when the ultimates appeared in the equations of their theories, the theories were no longer applicable. A new approach was needed.

Waves are probabilities

Radiation formerly conceived of as a continuum—an electromagnetic field—became particle-like photons. Electrons formerly thought of as particles also were discovered to behave as waves. Each phenomenon had both particle and wave properties—a paradox of major proportions.

All those beautiful differential equations of Maxwell describing electromagnetic behavior became more and more suspect. They worked very well in the macroscopic world, but broke down drastically in the microscopic confines of the atom (Figure 1-2).

The electron, always considered a particle, could be diffracted by a "grating" formed by a crystal lattice to produce a diffraction pattern, just like a wave. Electromagnetic waves shining on a photosensitive surface, although spread over a considerable area, caused a single electron to be emitted from a specific atom. The energy content

m = mass of the electron
−e = magnitude of the electron charge
Z = atomic number, charge of nucleus as
     multiple of electron charge magnitude
z = 1 for hydrogen atom
coulomb attraction = $e(eZ)/r^2$
centrifugal force = mv^2/r

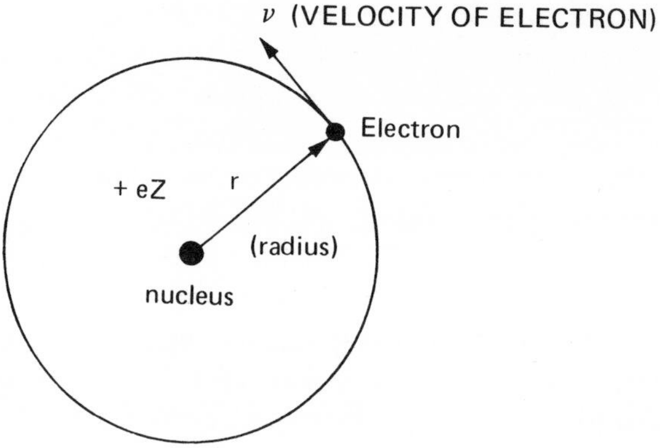

ν (VELOCITY OF ELECTRON)

Figure 1-2. The classical configuration of the atom merely balances the electrical Coulomb attraction against the mechanical Newton centrifugal force:

$$e^2Z/r^2 = mv^2/r.$$

Therefore the kinetic energy of the electron is:

$$\tfrac{1}{2}mv^2 = e^2Z/2r.$$

The classical theory fails because it doesn't explain why the electron does not radiate energy and fall into the nucleus, since a rotating electron (accelerative charge) must radiate according to Maxwell's equations. Also, classical theory gives no reason why an electron can have only certain restricted energy levels in the atom.

of this electron had the energy content of the spread-out beam. How can a single "solid" electron form a diffraction pattern? How can radiation energy spread out over a wide area be absorbed by a single electron?

A great number of theories were explored, but no amount of tampering with the classical theories based upon the continuum produced a sufficiently universal theory to account for these and the many other problems that kept appearing.

Bohr's "atomistic" quantum theory (Figure 1-3) replaces the idea that energy can be infinitesimally divided; however, his quantum mechanics are only an arbitrary set of rules that seem to work in predicting behavior at atomic dimensions and energy levels. This quan-

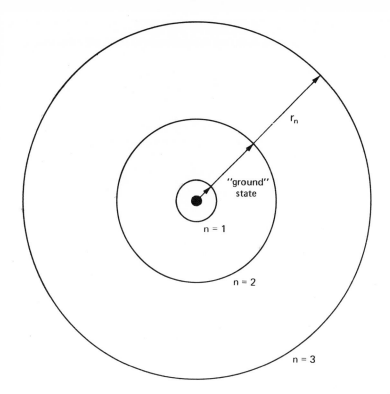

n = 1

"ground"
state

r_n

n = 2

n = 3

Bohr's hydrogen atom:

$r_n = 0.523 \times 10^{-8}\ n^2$ cm.

Only orbits that correspond to the integers n = 1, 2, 3 . . . are allowed. And radiation is absorbed or radiated only when electrons "jump" between "allowed" orbits.

Figure 1-3. Bohr's quantum mechanics provides a rather arbitrary set of rules that merely say that an atom does not radiate or absorb other than in discrete quantities $\Delta E = nh$, which corresponds to quantized angular momentum $nh/2\pi$, thus:

From Figure 1-2
$$e^2Z/r^2 = mv^2/r = mv^2r/r^2$$
$$= (mvr)\ (v/r^2)$$
$$= (nh/2\pi)\ (v/r^2),$$
$$\text{(angular momentum) } mvr = nh/2\pi,$$
$$\text{where } n = 1, 2, 3, \ldots$$

Since we know* h, m and e, and Z = 1, for the hydrogen atom at ground orbit, n = 1
$$n^2h^2/4\pi^2me^2 = 0.523 \times 10^{-8}\text{cm} = r.$$

Orbits then can exist only in multiples of n^2, and the ground, or binding, energy of hydrogen is:

$$E = \text{total energy (kinetic plus potential)}$$
$$E = -2\pi^2me^4/n^2h^2$$
$$= -13.58/n^2 \text{ electron volts.}$$

The energy levels for the hydrogen atom are shown in Figure 1-6 for different values of n.

$$*h = 6.63 \times 10^{-34} \text{ Joule-sec}$$
$$m = 9.11 \times 10^{-28} \text{ grams}$$
$$e = 1.60 \times 10^{-19} \text{ Coulombs}$$

tum theory, by itself, does not supply a satisfying philosophy to account for this unique behavior. The classical continuum theory also works, but only for large energy levels and large dimensions.

According to de Broglie's theory (Figure 1-4), it is possible to conceive of a particle as an energy ''pulse'' or wave packet of limited physical extension traveling through space, but a wave packet after refraction is dispersed over a considerable area. What happened to the particle? How do you account for the extended wave front of a light wave suddenly concentrating all its energy on a single electron to dislodge the electron from its atom? The continuum wave has no mechanism to account for dispersed energy to appear suddenly in a concentrated region. What then is the electromagnetic wave? These are very fundamental questions.

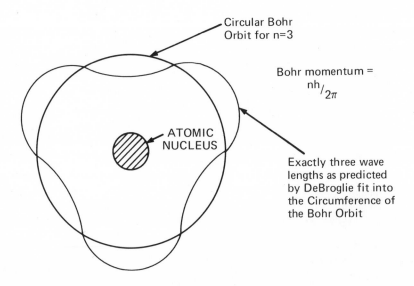

Figure 1-4. De Broglie postulated that a particle was a wave with length $\lambda = h/mv$ (particle momentum). For photons that have no mass, $\lambda = c/f$, in accordance with classical theory. Therefore a photon must have a momentum $= hf/c$. This theory only sharpened the wave-particle paradox. Schrödinger, by interpreting the wave as probability, went far to help resolve the conflict.

Scientists, until well into the twentieth century, were making a very basic assumption about the nature of the physical world that went largely unquestioned. The assumption was that continuous processes proceed from one state to the next in a perfectly deterministic way.

Starting with initial conditions, a process is preordained to proceed to the boundary conditions according to the rules of infinitesimals (a differential equation). There are no uncertainties and every step proceeds with no alternative choices possible. After all, don't the planets follow definite and predictable paths? When you place a pot of water on the fire, doesn't it always boil? It never freezes. Determinism fits in with Western man's predilection to command and control a definite future.

There are many areas that are not all that definite. Witness the very imprecise attempts at predicting the weather, economic conditions, the stock market, the course of politics and many other things.

In those situations, where very large numbers of factors determine a result, at best only a probable average and range of values can be predicted, but man's deterministic attitude tries to explain this away by claiming that a definite result can be predicted except that:

- Information is not complete.
- It would take too long.
- It would cost too much.
- The precise rules of behavior are not known.

Almost never is it felt that perhaps the physical world is inherently nondeterministic and that certainty is the exception. Perhaps even those things that appear so certain are really only the gross average of a huge number of individual probabilistic events, and each individual event can never be determined with certainty.

Suppose that in the world of atomic dimensions, probability rules. What we then see in our macroscopic world is merely the resulting average of the huge numbers of these individual atomic-level alternative trials, each probabilistically determined.

If the photon particle is the "true" carrier of electromagnetic energy, then the electromagnetic wave can be interpreted to correspond to a probability function describing the chances that a photon is present. By the same reasoning, with the electron undoubtedly a particle, the electron "wave" corresponds then to the probability of the presence of electrons.

Let's examine this startling idea in more detail. The fundamental points are:

- The basic physical phenomena are probabilistic, not deterministic as assumed by classical physicists.

- The rules of probability must have wavelike properties to account for the success of the wave theory in dealing with light.
- Changes in the condition of a particle do not flow smoothly (continuously) from one set of values to another, but in discrete jumps or steps.

In a dynamic universe, events are described by the states of matter changing from one condition of energy, location and time to another. Let us say that state of a particle of matter or a photon is represented by the letters E (energy), p (momentum), x (position) and t (time).

Going from one state to the next may be represented by the simple set of equations as follows:

$$E_2 = E_1 + \Delta E$$
$$p_2 = p_1 + \Delta p$$
$$x_2 = x_1 + \Delta x$$
$$t_2 = t_1 + \Delta t$$

The Greek symbol Δ (delta), here means a small change in the associated quantity.

In the continuum concept, when going from state 1 to state 2, a particle can pass through the intervals ΔE, Δp, Δx and Δt with absolute certainty, and these intervals can have any value from zero to infinity. The continuum idea, however, is only an assumption; though it is very difficult to get out of our minds, it is not a correct assumption because it is not justified by modern scientific discoveries.

All evidence points to the fact that the interval quantities ΔE, Δx and Δt are not traversed in continuous infinitesimal steps with absolute certainty, but that these quantities instead are traversed under probabilistic control as determined by the energy levels and geometry of the particle's location.

As a matter of fact, experiment shows that a pair of very simple but fundamental equations relate these Δ quantities to each other, thus:

$$\Delta E \Delta t = h$$
$$\Delta p \Delta x = h \text{ (where h is a constant)}$$

These equations are known as Heisenberg's uncertainty principle. If we consider the first equation, it can be interpreted to mean that the energy (E) at a given time (t) cannot be known to any greater degree of precision than the quantities ΔE and Δt, even with ''perfectly'' accurate measurements. Another way of saying this in terms of the second equation is—if the momentum is known to be within the interval Δp, then the position of the particle can't be determined any closer than Δx.

This interpretation provides us with the idea of indeterminancy as a prime principle. The idea of atomism versus the continuum is also brought out. Consider the Δ quantities as the state of transition mentioned before. It is easy to see that no change, Δ, can be zero, as required by the continuum concept, since then the other quantity paired with it would have to be infinite. Therefore you must conclude that each ΔE or Δp cannot be smaller than some finite, though very small, interval.

Thus we have almost answered the question of why there are energy gaps in atomic energy transitions.

The next step is to find out what this small but never zero Δ quantity is. Another rather simple experiment based on the photoelectric effect provides a direct answer. Classical theory would lead you to believe that electrons are expelled from the surface of a photosensitive surface with velocities (or energies) in direct proportion to the intensity of the light. Experiment proves otherwise (Figure 1-5). Instead the following relationship is found.

$$\text{Kinetic energy of emission} = nhf - \emptyset$$
$$(n = \text{integers } 1, 2, 3, \text{etc.})$$

where f is the frequency of the radiation striking the photo surface, h is a constant[2] named after its discoverer, Planck (in 1905), and $\emptyset$ is another constant called the ''work function'' which depends on the type of photosensitive material used. Work function is the amount of energy just sufficient to free an electron from the surface of the material with zero velocity. The energy of the emitted electrons does not change with the intensity of the light. More light generates more emitted electrons, but only the frequency of the light gives each electron more energy.

This unexpected result was explained by Einstein in 1905. An electron can have only discrete energies in steps equal to nf, 2hf, 3hf, etc. If n = 1, then hf is the smallest change ΔE that an electron can

---

[2]This is the same constant that appears in Heisenberg's equation on page 28 and in Bohr's relationship on page 25.

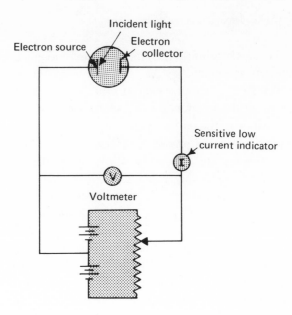

Incident light

Electron source

Electron collector

Sensitive low current indicator

Voltmeter

Figure 1-5. A plot of the voltage needed to stop current flow for different incident light frequencies provides remarkable results. Different photo materials provide curves of equal slope. Stopping voltage does not depend upon light intensity.

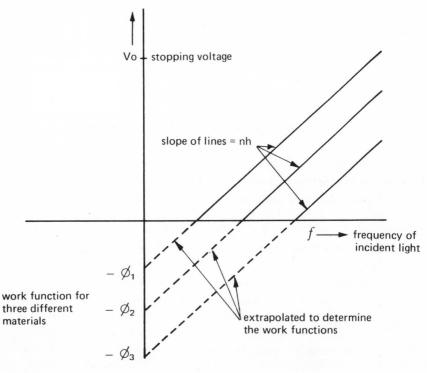

Vo — stopping voltage

slope of lines = nh

f ⟶ frequency of incident light

$-\phi_1$

$-\phi_2$

work function for three different materials

extrapolated to determine the work functions

$-\phi_3$

experience. This minimum energy step or "quantum jump" fits with the Heisenberg uncertainty principle.

Energy states of atoms and crystals

If electrons can change their energy states only in quantum jumps, the same principle must apply to atoms (Figure 1-6). Calculations made by Bohr (1913), de Broglie (1922) and Schrödinger (1926) fitted experimental data, especially for the simpler atoms such as hydrogen (Table 1-2). But what about the complex structure of a crystal lattice of many atoms?

The possible energy states of an atom can be resolved into discrete lines (frequencies), with gaps in between. But when many atoms are in close proximity, the simple line structure is modified to densely packed lines, covering a broad band of frequencies. As in Figure 1-1, one broad band of energy levels corresponds to the electrons held close to the atom centers (nuclei), and a second band belongs to the group of electrons that may have escaped from the grip of the nuclei and are free to drift among them.

Should these two bands overlap, then the material is a conductor. In a conductor electrons can easily cross between being held by the nuclei (called valence electrons) and being able to drift free (called conduction electrons). A wide gap between the conduction and valence-energy bands makes the material into an insulator. In insulators a large amount of energy is required to bridge the energy gap and pull valence electrons into the conduction bands. A medium-energy gap creates a semiconductor, the type of material that is one of the subjects of this book.

Atomic structure and valence

In a simplified concept of the structure of the atom, the atom is considered to consist of a nucleus that contains most of the mass and all of the positive charge. The nucleus is surrounded by electrons in orbits. It is thus a sort of solar system in miniature.

Each atomic nucleus contains as many protons as its atomic number, together with enough neutrons to add up to the mass number. Since each atom is electrically neutral, there are as many electrons as protons. These electrons are arranged around the nucleus in definite energy levels or shells.

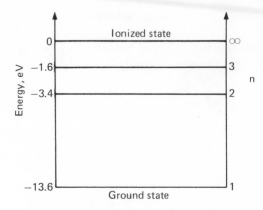

Simplified energy level diagram for
the Hydrogen atom.

One electron-volt = 1.6×10^{-19} joules.

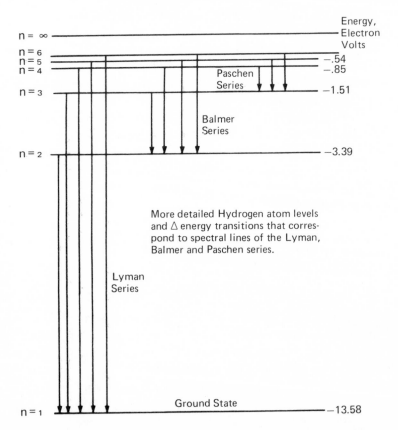

Figure 1-6. Energy levels of the hydrogen atom. Energy levels are permitted to change only between allowed states that correspond to n = an integer, when absorbing or radiating energy.

32

Period	Series	I	II	III	IV	V	VI	VII	VIII	0
1	1	1 H 1.00797								2 He 4.0026
2	2	3 Li 6.939	4 Be 9.0122	5 B 10.811	6 C 12.01115	7 N 14.0067	8 O 15.9994	9 F 18.9984		10 Ne 20.183
3	3	11 Na 22.9898	12 Mg 24.312	13 Al 26.9815	14 Si 28.086	15 P 30.9738	16 S 32.064	17 Cl 35.453		18 A 39.948
4	4	19 K 39.102	20 Ca 40.08	21 Sc 44.956	22 Ti 47.90	23 V 50.942	24 Cr 51.996	25 Mn 54.9380	26 Fe 55.847 27 Co 58.9332 28 Ni 58.71	
	5	29 Cu 63.54	30 Zn 65.37	31 Ga 69.72	32 Ge 72.59	33 As 74.9216	34 Se 78.96	35 Br 79.909		36 Kr 83.80
5	6	37 Rb 85.47	38 Sr 87.62	39 Y 88.905	40 Zr 91.22	41 Nb 92.906	42 Mo 95.94	43 Tc [99]	44 Ru 101.07 45 Rh 102.905 46 Pd 106.4	
	7	47 Ag 107.870	48 Cd 112.40	49 In 114.82	50 Sn 118.69	51 Sb 121.75	52 Te 127.60	53 I 126.9044		54 Xe 131.30
6	8	55 Cs 132.905	56 Ba 137.34	57-71 Lanthanide series a	72 Hf 178.49	73 Ta 180.948	74 W 183.85	75 Re 186.2	76 Os 190.2 77 Ir 192.2 78 Pt 195.09	
	9	79 Au 196.967	80 Hg 200.59	81 Tl 204.37	82 Pb 207.19	83 Bi 208.980	84 Po [210]	85 At [210]		86 Rn [222]
7	10	87 Fr [223]	88 Ra [226]	89 Actinide series b						

a. Lanthanide series:

57 La 138.91	58 Ce 140.12	59 Pr 140.907	60 Nd 144.24	61 Pm [145]	62 Sm 150.35	63 Eu 151.96	64 Gd 157.25	65 Tb 158.924	66 Dy 162.50	67 Ho 164.930	68 Er 167.26	69 Tm 168.934	70 Yb 173.04	71 Lu 174.97

b. Actinide series:

89 Ac [227]	90 Th 232.038	91 Pa [231]	92 U 238.04	93 Np [237]	94 Pu [242]	95 Am [243]	96 Cm [245]	97 Bk [249]	98 Cf [249]	99 Es [254]	100 Fm [252]	101 Md [256]	102 [254]	103 Lw [257]

Example:

atomic number → 1H ← element symbol (hydrogen)

1.00797 ← atomic weight

Note: 1. Atomic weights are relative to the weight of the carbon isotope C^{12}, which has an atomic weight of exactly 12 units.

2. Bracketed atomic weights are radioactive elements.

3. If the atomic weight is rounded off to the nearest whole number, the mass number of the principle isotope of the element is obtained in most cases.

Table 1-2. Periodic Table of the Elements

Valence is a measure of the combining capacity of an atom when it forms molecules or crystal structures with other atoms. It can be partially explained on the basis of the configuration of the electrons in an atom's outermost shell. Valence bonds between atoms may be the result of:

- The transfer of one or more electrons from one atom to others to form ions (charged atoms).
- The sharing of electron pairs by atoms.

The first type of valence is called electrovalence, where the atoms are held together by electrostatic forces. The second is known as covalence.

For example, Group I, alkaline earths (Table 1-2)—Li, Na and K—are definitely electrovalent. They readily give up their one outer-shell electron to form positive ions.

Group VII, halogens—Fl, Cl, Br and I—readily accept a single electron to complete their outer electron rings and form negative ions.

The alkalines thus have a valence of $+1$ and the halogens a valence of -1. Together they form ionic compounds, which assume crystaline structures like the cubic arrangement of common table salt, NaCl.

However, an atom like carbon has four electrons in its outer shell. It can share four more electrons supplied by other atoms to complete its outer ring, or lend its four atoms and share them with, say, atoms of oxygen. Two atoms of oxygen each need two electrons to fill their outer shells; thus one carbon atom can form covalent bonds with two oxygen atoms to form nonionic carbon dioxide (CO_2). Carbon has a valence of ±4 and oxygen, -2.

This description of valence is very simplified. The field of chemistry contains many complex valence relationships, especially in covalent molecules as found in organic chemicals. However, for the purpose of understanding semiconductors, it is sufficient to associate the term valence electrons with those electrons in an atom's outer ring that enter into the formation of chemical bonds. These electrons are the easiest to separate from the atom.

Table 1-1 lists the energy in electron-volts (eV) needed to remove just the most loosely bound of the valence electrons. It is called the First Ionization Potential.

Chapter 2

How Semiconductors Are Basic To Solid-State Devices

A semiconductor is a material that conducts electrical current rather poorly compared to a metal. Pure elements that are semiconductors are called *intrinsic* semiconductors. The more familiar types belong to the carbon family of elements—those that have four valence electrons in their outer shell (Figure 2-1, column IV) and in Table 1-2. The elements—carbon, silicon and germanium—form the most important ones in this category.

Figure 2-1 also lists the boron and nitrogen groups, which are used in creating *extrinsic* semiconductors from intrinsic types, as explained later. This type of semiconductor is the kind most important to solid-state devices. The addition of columns III and V impurities to intrinsic materials in minute, controlled amounts determines the important properties of extrinsic semiconductors. Finally, a large class of materials, in use and under development called intermetallic compounds, are also considered semiconductors. These materials include such compounds as indium arsenide, indium antimonide, bismuth telluride, gallium arsenide, gallium phosphide and others.

Semiconductors do more than just conduct

Conductivity in a pure semiconductor can be changed drastically by many things. The most dramatic change occurs when you introduce impurities as small as one part per million to replace some of the atoms of a pure semiconductor. The resulting extrinsic material can have a conductivity that is several orders of magnitude larger than the original

III	IV	V
Boron B(5)	Carbon C(6)	Nitrogen N(7)
Aluminum Al(13)	Silicon Si(14)	Phosphorus P(15)
Gallium Ga(31)	Germanium Ge(32)	Arsenic As(33)
Indium In(49)	Tin Sn(50)	Antimony Sb(51)
Thallium Tl(81)	Lead Pb(82)	Bismuth Bi(83)

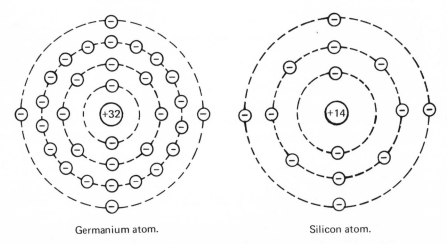

Germanium atom.   Silicon atom.

Figure 2-1. The elements listed in the three columns are groupings from the chemists' Periodic Table of the elements most important to the manufacture of solid-state devices. The numbers in the parentheses are the atomic numbers of the elements and the column headings III, IV and V refer to the valence numbers of the elements in the column. Silicon and germanium both have a valence of four, since their outer electron shells contain four electrons which can enter into chemical bonds with other elements.

intrinsic material, and vastly increased sensitivity to a wide variety of external influences. Temperature can cause a highly exponential increase in conductivity. Shining light on some materials can cause orders of magnitude change in conductivity. Even magnetic fields have strong influences in some materials.

At first researchers were unable to control and predict these dramatic effects; now most of these strong dependencies of conductivity on external influences can be predicted and controlled. The result? Semiconductors have become the basis for innumerable useful devices.

What is the conduction mechanism of semiconductors that leads to these remarkable properties?

Unlocking the mysteries of semiconductors

Conduction in a metal results from the many "free" electrons, about 10^{23} per cm^3, that drift under the influence of an applied voltage. In a semiconductor, free conduction electrons are scarce. Unlike metals, semiconductors have another type of current carrier called *holes*. These holes behave as if they were positive charge carriers. In intrinsic semiconductors they contribute as much to current flow as the electrons do, and in extrinsic semiconductors, they can contribute a great deal more. Thus the total current flow in a semiconductor is determined by both electron and hole drift.

This drift is an average motion directed by an applied voltage that is superimposed on primarily random motion. The random motion results from thermal energy.[1] Even though current flows, the net charge within the material (metals or semiconductors) always remains zero, since the electrons are balanced by the positive charges of the "immobile" atomic nuclei; the negative-voltage source adds electrons as other electrons are drawn off at the positive terminal.

Though the nuclei in solids don't normally flow, their "immobility" doesn't prevent them from vibrating about fixed points. All materials at zero degrees absolute (Kelvin) have a minimum of thermal energy and the nuclei vibrate very little; also, the materials' electrons are in the lowest possible energy states. At higher temperatures, the material acquires thermal energy from its surroundings. Most of this thermal energy takes the form of vibrations of the nuclei about equilibrium positions. Electrons can absorb some of this energy only when the nuclei vibration levels reach above certain thresholds which correspond to electron allowed energy bands. When electrons are on the average absorbing energy at the same rate as they are giving it back to the nuclei, the electrons are said to be in thermal equilibrium with the material.

---

[1] See page 70 for energy equation.

In semiconductors at room temperature (300°K), a few nuclei have enough energy to kick some electrons from the valence band into the conduction band. This must be done in one step since quantum mechanics forbids an electron to gradually accumulate energy. In insulators, however, many fewer nuclei have enough energy to cause electrons to cross the wider forbidden gap, but the distinction between insulator and semiconductor is only a matter of degree.

The forbidden energy gaps of two common semiconductors are 0.7 eV for germanium and 1.1 eV for silicon. Diamond, which is an insulator, has an energy gap of about 7 eV (eV is energy measured in electron volts).

Holes conduct electricity too

When an electron passes into the conduction band, it leaves behind a vacant energy state in the valence band. This vacancy is the so-called *hole*. In pure or intrinsic semiconductors, electrons in the conduction band and the holes in the valence band occur in pairs. The number of such pairs is small compared to the number of nuclei, and their quantity decreases both as the forbidden energy gap increases and as the material's temperature decreases. And when the number of pairs is very small at normal temperatures, the material is an insulator.

Once an electron passes into the conduction band, the electron can then absorb energy in an almost continuous fashion, because here there are many closely-spaced energy levels that the electron can occupy. At the same time, as the electron is pushed to a higher energy by, say, an applied electric field, the hole that was left behind is forced deeper into the valence band. The valence band also has many closely-spaced levels.

In the absence of electric fields, the conduction electrons and holes move about the material in a purely random fashion. Their directions change every time they encounter a vibrating nucleus and there is an exchange of energy. Figure 2-2 shows a typical random path. Between collisions with the nuclei, the electrons and holes obey all the laws of motion of free particles. Thus when an electric field is applied to the material, the particles are accelerated. When the accelerated particles again strike nuclei, they can pass energy gained from the field to the nuclei. This results in heating of the material. The average of all the accelerated motions is the particle flow or current; this depends

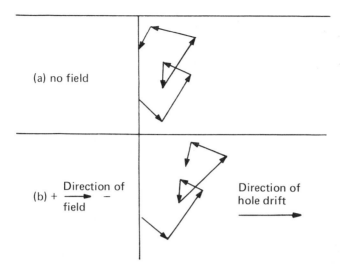

(a) no field

(b) + →— field

Direction of field

Direction of hole drift

→

Figure 2-2. The random motion of a current-carrier particle takes on an added drift under the influence of an electric field.

upon the number of free particles and the average velocity that they gain between collisions.

Crystal lattices hold together with covalent bonds

The nuclei of semiconductors in the pure state are arranged in symmetrical patterns called crystalline structures. In crystals of germanium and silicon, covalent bonds form between adjacent nuclei (Figure 2-3(a)). Each atom shares an electron with each of its four neighboring atoms and therefore each nucleus is influenced by eight electrons. This sharing results in each individual atom behaving as if it had a complete subshell of eight electrons.

In an intrinsic semiconductor, when one of the covalent bonds is disrupted by the loss of one of the shared electrons, a vacant place or hole is left behind in the bond and an electron-hole pair is formed (Figure 2-3(b)). Since the hole tends to attract and hold electrons, an electron that had been released from another nearby bond can be captured and held to again complete the bond. Thus, the hole appears as if it had moved from one position to the other, as an electron dislodged from one bond is captured by another vacant bond. This apparent

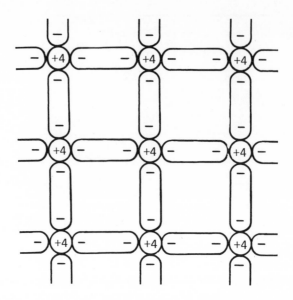

Figure 2-3(a). Each atom of silicon or germanium in a crystal lattice shares one electron with each of its four neighbors to form covalent bonds.

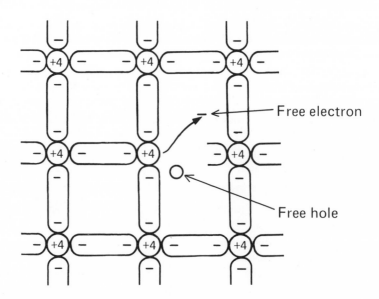

Figure 2-3(b). A disrupted covalent bond produces a hole-electron pair.

motion of the holes is in the opposite direction from the flow of the electrons, and both the "flow" of holes and electrons add to contribute to the total electric current since holes behave as if they were positive charges.

Impurities make the difference

Intrinsic or pure semiconductor materials have few applications in electronics. Extrinsic semiconductors, though, make up the bulk of the material used to manufacture modern solid-state devices. Carefully selected impurities are deliberately added to pure semiconductor materials. The type, quantity and location of these impurities determine most of the semiconductors' useful properties.

First very pure semiconductor material, usually germanium or silicon, is produced in special furnaces (Figure 7-1). Then impurity, or doping, materials are introduced at specific locations on the pure host material. Refer again to Figure 2-1, which is a portion of a periodic table of elements. The column labeled IV contains elements that have four electrons in the outer (valence) atomic shell, and includes germanium and silicon. Dopants are selected from columns III or V, which contain three and five electrons in their valence shells. The amount of doping materials that is used can be as small as one impurity atom for 10^8 host atoms.

Impurity atoms that readily fit into the host's crystal lattice to replace host atoms are called *substitutional impurities*. Other impurities which can't fit into the lattice structure and take up positions between the host's atomic nuclei are called *interstitial impurities*. We will only consider the substitutional types, which are the most important dopants in solid-state electronics.

Let's examine the effects of impurities from column V such as arsenic or antimony. An impurity with a valence of five that is surrounded by a host material with a valence of four has an extra electron left over after it has formed covalent bonds with its host neighbors (Figure 2-4(a)). This extra electron is excluded from the valence band of energies. Since all the bonds are complete, neither can it exist in the forbidden energy gap, but a little extra energy, readily picked up from thermal agitation, can put it into one of the many available conduction bands. Column V impurities thus increase the number of free conduction electrons in a semiconductor crystal. Such dopants are therefore

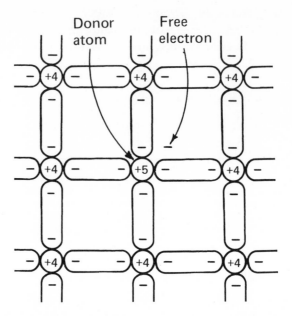

Figure 2-4(a). An n-type semiconductor, formed by using a valence-5 donor impurity, provides a "free" conduction electron.

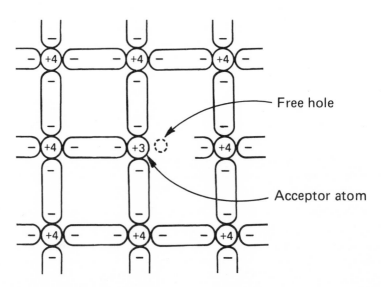

Figure 2-4(b). A p-type semiconductor is formed when a valence-3 acceptor impurity fills one of the crystal-lattice locations. The lack of an electron to make a covalent bond creates a "hole," which behaves as a current carrier.

called *donor* impurities. Since the donor electron in the conduction band is "free" to roam and leave its nucleus, the donor-impurity atom becomes an ion with a net localized positive charge. Overall, though, the host material is electrically neutral.

Extrinsic semiconductor materials that are doped with extra electrons, or donor impurities, are known as *n-types*, because of their contribution of negative (electron) charge carriers.

If, on the other hand, an impurity is selected from column III, instead of providing an extra electron, the impurity creates a hole (Figure 2-4(b)). An additional electron is needed to complete one of the dopant's covalent bonds. The impurity is then called an *acceptor*, because it can accept an electron from a nearby host atom. When an electron is captured and temporarily held, the impurity atom exists as a localized negatively charged ion, though again, the material is electrically neutral overall. Now, since "free" holes have been introduced into the host material, the material is called *p-type* material, since holes act like positive charges. Where in n-type material the electron current carriers acquire conduction band energy levels, in p-type material the holes assume valence band energies (Figure 2-5).

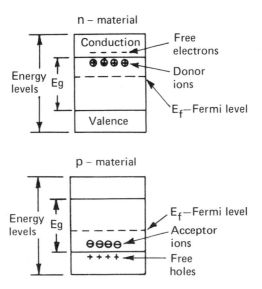

Figure 2-5. Relative energy levels at room temperatures for n- and p-type semiconductor materials. Note that the average electron energy—the Fermi level E_f—is lower in p-type materials.

The majority carrier rules

Every semiconductor has some degree of each type of charge carrier, but the larger quantity of the two types is called the *majority carrier*. The small quantity is the *minority carrier*. Electrons are the majority carrier in n-type material and holes are the majority carrier in p-type materials. In the more useful semiconductor materials, at room temperatures the number of majority carriers is determined mostly by the amount of impurity. The majority carrier count is usually made to be high as compared to the number of carriers supplied by the semiconductor's intrinsic properties. The material then has the desirable property that the over-all carrier concentration remains almost constant with temperature. The temperature range over which the carrier concentration, as measured by conductivity, stays constant is called the *extrinsic range* (Figure 2-6). However, note that at high temperatures (above 150°C) the number of thermally, or intrinsically, generated hole and electron pairs becomes large enough to swamp out the extrinsic carriers. The conductivity then takes on a marked temperature dependence, and the crystal again becomes intrinsic. This higher temperature range is called the *intrinsic range*.

The temperature at which the crystal goes from extrinsic to intrinsic behavior is a very important specification. This temperature sets the upper limit at which devices made from the material are able to operate.

There is also a very low temperature at which the majority carriers begin to "freeze" into the impurity atoms and again, desired extrinsic behavior ceases. This temperature point, however, is usually too low to be of importance to most common solid-state device applications.

Fields and light affect semiconductors

When electric fields applied to semiconductors are moderate (less that 10^5 V/cm for Ge at room temperature), the energy that a carrier can pick up between collisions in its free-path movement is not enough to do more than impart thermal agitation upon colliding with a nucleus. Above a threshold field strength, though, an *avalanche* process occurs. Above the threshold, the charge carrier can acquire enough energy to dislodge an electron from a nucleus to create a new electron-hole pair. This new pair, in turn, can produce another pair, and so on. Thus the

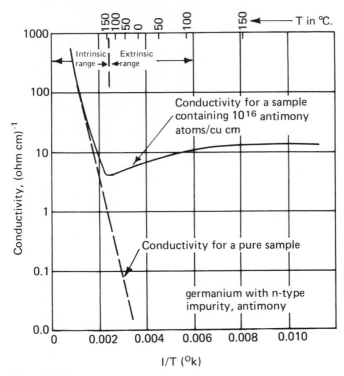

Figure 2-6. At high temperatures the intrinsic thermally generated electron-hole pairs swamp the extrinsic majority carriers produced by the dopant, and the material's conductivity becomes markedly temperature-dependent.

avalanche that results can rapidly raise the current flow enormously. Under these circumstances, the material is said to have *broken down*.

The field strength necessary for breakdown increases with an increasing forbidden energy gap and also changes with temperature.

Light shining on the crystal can also generate carrier pairs. However the wavelength of the light must be short enough so that the wave's photon energy content can overcome the material's forbidden energy gap. Light with wavelengths longer than this merely passes through the crystal, as through a window; thus materials with high forbidden energy gaps, like diamond, tend to be transparent. A low energy-gap material, germanium, with only a 0.72 electron-volt gap, is transparent only for very long wavelengths—greater than 17,000 Å which lie in the infrared band.

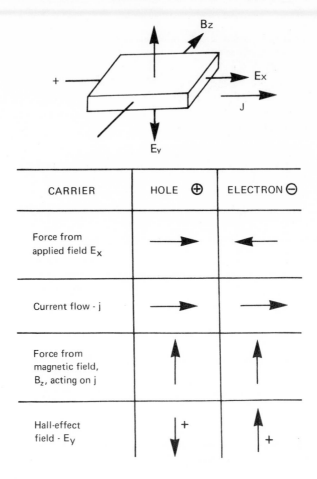

CARRIER	HOLE ⊕	ELECTRON ⊖
Force from applied field E_x	→	←
Current flow - j	→	→
Force from magnetic field, B_z, acting on j	↑	↑
Hall-effect field - E_y	↓ +	↑ +

Figure 2-7. A Hall-effect field results when a magnetic field is placed at right angles to an applied electric field. The direction of the Hall field is a direct indication of the type of majority current carrier in the material.

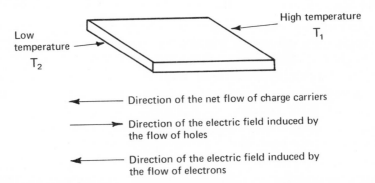

Figure 2-8. A temperature gradient in a semiconductor crystal will induce a net drift of charge carriers to the cooler end. The direction of the induced field indicates the majority current carrier in the material.

The production of both field-induced and light-induced current-carrier pairs will be explored more fully later. Obviously, sensitivity to light finds innumerable applications and the avalanche effect also has its uses and dangers in practical devices.

A magnetic field can tell them apart

The concept of hole or p-type semiconductor conductivity is more than just a convenient representation. It is very real; many experiments clearly show the difference between p- and n-type materials. If not for the hole theory, these experimental results would be difficult to explain.

The most widely known experiment is the Hall effect. If a magnetic field is placed at right angles to an electric field, which is applied across a semiconductor sample, a force will act on the charge carriers in the sample, as shown in Figure 2-7. If the charge carriers are holes, the current drift tends to be diverted so that a small downward electric field is created. The top surface of the material assumes a positive charge and the bottom, negative. In n-type material the sample surfaces are charged in the opposite direction.

This very real and easily measured effect becomes quite pronounced in materials like antimony and bismuth. The Hall effect, in addition to revealing the type of majority carriers a material sample has, also allows measuring their mobility and concentration (number/unit volume). And besides serving as an important tool for the study of solid-state theory, the effect is the basis for very useful and practical devices. Several applications will be explored later.

A similar effect can be obtained by the use of heat. If one end of a crystal is heated, the charge carriers at the hot end have a higher mean speed between collisions than those at the cold end. Thus there will be a net drift of carriers towards the cold end. This drift produces a potential difference between the two ends of the crystal. The sign of the electric field produced by this potential difference depends upon the type of majority carrier in the material (Figure 2-8).

Chapter 3

PN Junctions: the Keys to Understanding Solid-State Devices

Place bulk semiconductor material between two large metal plates and make good physical contact; then run electric current through this sandwich. You probably will find that current flows equally well in either direction between the plates. Now substitute a small metallic point for one of the plates. There is a good chance that current may flow better in one direction than the other. This rectification action has been known for many years (since about 1902) and was the basis for the so-called ''cat's whisker'' crystal detector (Figure 3-1(a)) of early radio days (1906 to 1920).

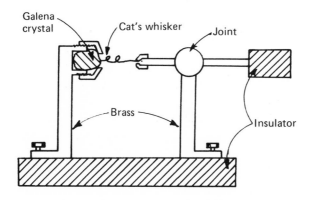

Figure 3-1(a). A crystal detector, used in early radio communications, often required a tedious search for a "sensitive" spot on the crystal where efficient rectification could be obtained.

How the point contact produces rectification has been the subject of much heated discussion in the literature of solid-state physics, and many theories have been proposed. In addition, the manufacture of point contact devices was difficult to control, and those that were produced proved to be rather unstable performers.

Work on both the theory and practice of crystal detectors soon died away when radio tubes almost completely replaced the crystal in radio communications about 1925. The fine wire of a crystal detector's point contact required frequent adjustment, and finding a new sensitive spot was tedious. The diode tube was a lot more stable, and after the third, or "grid," element was added, the tube also provided amplification.

However, two sandwich-type semiconductor devices which were developed about 1926—the copper-oxide and selenium rectifiers—maintained their position as power rectifiers for many years, and the selenium units are still used to this day.

Tubes were noisy at high frequencies, however, and had other undesirable qualities. Since the very important World War II radar systems relied heavily on mixer and detector circuits that could be implemented with diodes, a great deal of research went into developing an improved solid-state detector for high frequencies. Silicon and later germanium were found to provide the desired characteristics in what was at that time considered high-purity (99.9%) materials. When controlled amounts of specific impurities were added, improved performance was obtained. By 1948, before the transistor was announced, semiconductor devices made up a small but important and specialized part of the electronics market. Selenium rectifiers were widely used in electronic power supplies and silicon and germanium diodes filled an important niche in the microwave field, but the 1948 diode (Figure 3-1(b)) differed little from the 1920 cat's whisker detector rectifier. The effects of purity, doping and surface preparation were still not well understood. Manufacture of these devices was more an art, based upon empirical experience, rather than a science.

In 1948 the addition of another point contact to a point-contact diode by Drs. Shockley, Bardeen and Brattain at the Bell Laboratories, gave birth to the point-contactor transistor. Even though, as Shockley likes to point out, the point-contact transistor was not a chance discovery but the result of Bardeen's theory of surfaces, the manufacture of

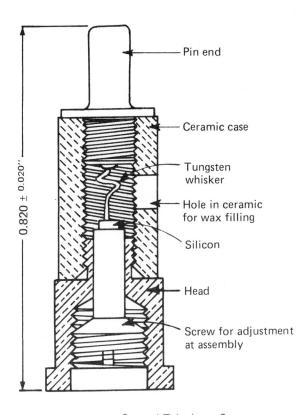

0.820 ± 0.020″

Pin end

Ceramic case

Tungsten whisker

Hole in ceramic for wax filling

Silicon

Head

Screw for adjustment at assembly

General Telephone & Electronics

Figure 3-1(b). By 1948, efficient fixed crystal diodes were widely used in high-frequency work such as for mixing circuits in radar receivers.

point-contact transistors based on this theory proved to be difficult, and the units highly variable and unstable.

Disputes about surface phenomena soon became academic because in mid-1949, a series of papers in the Bell System Technical Journal, in particular Shockley's paper, presented a new concept—the pn junction—which included a theoretical treatment of the pnp transistor. Unlike point-contact devices, these new junction devices depended upon the internal crystal structure of high-purity materials, rather than surface phenomena.

Ohmic junctions don't rectify

Consider a single crystal of germanium or silicon doped so that half the material is p-type and half n-type (Figure 3-2). The boundary formed between these two material types is called a pn junction. As we shall see, this type of junction can provide diode action. Some point-contact diodes work because the point touches a region in the material where a pn junction had been fortuitously or purposely formed. A junction of a metal and semiconductor also can be made to rectify, as in the case of selenium and copper-oxide rectifiers. Some point-contact diodes depend on metal-semiconductor action for rectification, but in most instances metal-to-semiconductor junctions as used in modern solid-state devices are intended to behave as ohmic contacts.

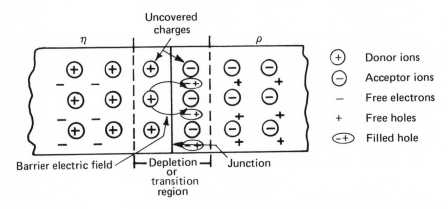

Figure 3-2. The boundary region between a p- and n-type material becomes depleted of current carriers and a barrier field develops across the junction. The size of this region for germanium and silicon varies between roughly 10^{-4} and 10^{-5} cm.

An ohmic junction, by definition, conducts electricity equally well in either direction. This means that the junction region between the materials does not have a net directional preference in the transport of current carriers across the interface. In large area contacts, localized directional barriers that may form are likely to be distributed randomly and thus produce a net cancellation of any directional bias to current flow. This is especially true when materials are joined by a heating process such as soldering or welding, since heating tends to disrupt the orderly lattice structure of a semiconductor material and with it any tendency towards current directionality. If bulk or powdered semiconductor material is placed between large conducting surfaces, the crystal orientation in the contacted material is likely to be sufficiently random to cancel any overall directional effect. Even with intact crystal structures, metal-to-semiconductor junctions can be made ohmic by the proper selection of materials, as described later.

If a rectification action is what you desire, the modern approach is a deliberately made pn junction. It can provide much more stability and reproducibility than any point-contact device. Its efficiency and other desirable properties as a diode for circuit design exceed metal-to-semiconductor sandwich devices.

A pn junction forms a space-charge region

When p and n materials are initially placed into contact, there is a brief flow of current across the interface boundary (Figure 3-2). Electrons flow from the n side to the p side and holes flow from the p side to the n side. This leaves a region in the vicinity of the boundary between the two materials that is depleted of current carriers—a so-called uncovered area where impurity ions face each other across the interface with their charges no longer neutralized. This boundary is also often called the transition or space-charge region.

The flow or diffusion of these carriers across the junction thus leaves the region on the n side of the junction with a positive charge and the p side with a negative charge. As in a charged capacitor, an electric field connects the charges across the junction. After the initial flow of carriers and buildup of the electric field, the transition region reaches a state of equilibrium and current surge ceases.

In the last chapter we learned that the average energy of electrons in n-type material is higher than that in p-type material; thus the valence-conduction electrons on the n side will initially, on the average, tend to flow over to the p side at a greater rate. It should be noted that a flow of valence electrons from n to p sides is equivalent to holes flowing from p to n. This flow soon ceases when the average electron energy levels (called Fermi levels E_f) in the two contacting materials equalize. Figure 3-3 shows the two energy level diagrams of Figure 2-6 for p and n materials aligned side by side so that their E_f levels match. This is the energy diagram for an unbiased pn junction. As might be

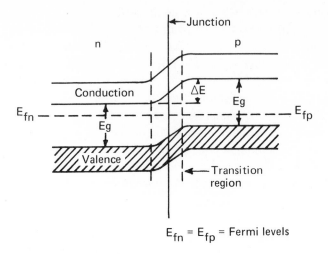

Figure 3-3. Barrier energy ΔE is the minimum energy an electron must have to be able to travel from the n to the p side once equilibrium between the two materials is established. Typical ΔE values for germanium range from 0.3 to 0.6 volts and about 0.5 to 0.8 volts for silicon.

expected from the law of conservation of energy, energy equilibrium between the two materials must automatically adjust so that net current flow is zero in the absence of an external energy source.

Passing current through the junction

The interesting point in this whole discussion of pn junctions is that an externally applied voltage encounters different "resistance" to the flow of current which depends upon the direction of the voltage.

Let a small voltage E_D be applied across a pn junction as in Figure 3-4(b). Electrons immediately will flow from the battery through the metallic contacts to the n region while holes flow into the p region. Holes flowing into the p region are the equivalent of valence electrons flowing out of the p region. When the electrons and holes reach the junction, the potential barrier slows them down and causes a "pile-up" in the transition region so that more electrons and holes occupy this region than are found there in the unbiased equilibrium condition. Depletion is reduced in the transition region and its size is reduced. In this way the electrostatic field across the junction drops

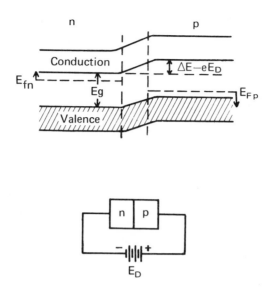

Figure 3-4. In forward-biased pn junctions the barrier energy ΔE is reduced by $e \cdot E_D$ to offer less opposition to the passage of conduction electrons.

and the energy an electron must have to jump the barrier also becomes less by an amount proportional to the applied voltage, $e \cdot E_D$, where e is the charge on an electron. This is called a forward-biased junction. The relative energy relationships are shown in Figure 3-4(a).

Reverse current is small

Reverse the direction of E_D and now the junction is reverse-biased (Figure 3-5). Electrons will try to flow out of the n material and holes out of the p material. Instead of tending to pile up current carriers at the junction and reduce the space charge, the transition region now becomes depleted of majority carriers and the space charge increases. The transition region grows larger and the barrier potential across the junction increases. The barrier, in a very short time, becomes almost equal to the applied potential and current flow almost ceases. Only a very small percentage of the majority carriers on each side of the junction have sufficient kinetic energy to make it over the increased potential gap.

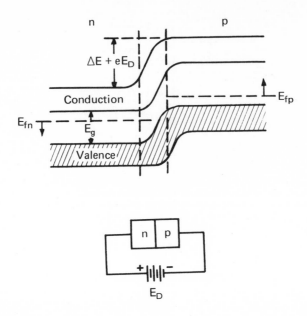

Figure 3-5. Reverse-biased pn junctions increase the barrier-energy step and oppose the flow of current.

What little current does flow is due to the intrinsic generation of electron-hole pairs in the undoped portions of the semiconductor materials. These pairs constitute the minority carriers. In the discussion on pn junctions, we have ignored the actions of the minority carriers thus far because their presence in the forward-biased case is overshadowed by the much larger majority-carrier effect at room temperatures. The reverse current is very nearly independent of the reverse-bias voltage and depends mostly on electron-hole pair generation. And since pair generation is mainly the result of thermal agitation, the reverse current, often called reverse saturation current, is temperature dependent. In fact, when temperature rises too high (Figure 2-7), the intrinsic properties of the material take over and characteristics dependent upon majority carriers become hidden.

Diode action of pn junctions

At room temperature, the saturation region current is almost independent of applied voltage (Figure 3-6). Since the forward currents are so much greater than the reverse currents, the junction acts as a rec-

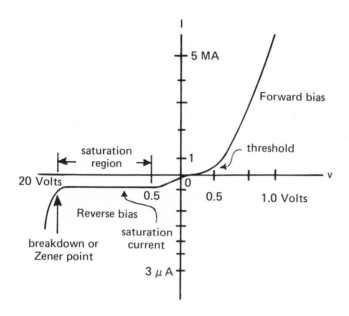

Figure 3-6. Note that the forward and reverse bias voltage scales are not in proportion. The ratio between forward current at 0.2 V and reverse saturation current at up to 20
V can be 2000:1 or higher in a good silicon diode.

tifier. The ratio of forward to reverse current for a good semiconductor
diode at room temperature is about 50 to 1 for an applied voltage of 0.1
V and rapidly increases to over 2000 to 1 above 0.2 V.

The reverse-current saturation region is not absolutely constant
with voltage, nor as constant as the theory predicts, but a well-made
diode can be made with reverse current flat to a few percent. The slope
of this curve is determined more by the empirical manufacturing treatment than any theoretically established criteria. This saturation region
of the pn junction has an important application, such as a variable
capacitor, discussed later. Beyond the breakdown, or Zener point, the
pn junction is used in power supply regulator circuits. Diodes used for
this purpose, known as Zener diodes, will also be discussed in greater
detail later.

The sensitivity of the reverse current to temperature can be illus-
trated by the fact that in a germanium junction the reverse current
increases about 10 percent for each one-degree increase in temperature. Silicon provides an even greater percentage change; however, its

absolute current value is much lower at any given temperature. The reverse, or saturation, current for a germanium 5-Ω-cm sample is roughly 10^{-3} A/cm^2 and 10^{-6} A/cm^2 for a silicon sample at room temperature.

Since the transition region in the neighborhood of the junction is only about 10^{-4} cm thick, a small voltage develops a sizeable electrostatic field. Electron-hole pairs that move through the junction under the influence of such a high field can pick up enough kinetic energy to create another pair. Above a certain voltage, the rate of pair generation suddenly increases enormously. This critical voltage is known as the breakdown voltage or Zener point. The voltage needed for breakdown depends upon the impurity concentration of the n and p regions and the temperature. A lower impurity concentration produces a lower breakdown voltage.

Capacitive action of pn junctions

Whan a pn junction is reverse biased, the depletion region becomes wider because more current carriers are forced away from the region. The width of the depleted region increases with the amount of back voltage that is applied. Because the depleted region is swept clear of "free" current carriers, the material in it behaves as a dielectric in a capacitor. But it is a capacitor whose capacitance is an inverse function of the applied back voltage. Some materials provide a capacitance inversely proportional to the square root of the voltage; others may vary inversely as the cube root.

Some diodes are specially designed to exploit this property of a pn junction. Commercial units, called varactors, varicaps or simply variable-capacitor diodes, can be made to change their capacitance from a few picofarads to several hundred. Microcircuits fabricated with so-called monolithic techniques, or integrated circuits, frequently use pn junctions as capacitors, but usually with fixed capacitor values. More information on integrated-capacitor diodes will be provided in Chapter 7.

When a pn junction is forward biased, a different kind of capacitor action results. The "pile-up" of charges across the junction, which was previously mentioned, can be represented by an equivalent capacitor, called diffusion capacitance, that stores the same amount of charge. This stored charge can be quite large and, in switching applications, diffusion capacitance can impose severe limitations on a high-speed circuit's reverse recovery time.

Effects of light on junctions

Reverse-current carriers in pn junctions come mainly from electron-hole pairs that are generated from thermal-energy agitation within the diffusion, or depletion, region; however, if the junction is arranged so that it can be irradiated with light of the proper frequency (Figure 3-7), additional carrier pairs can be generated. Like the thermally-generated carrier pairs, the pairs that result from light also tend to "discharge" the barrier gap and reduce the electrostatic field that bridges the junction. Under influence of the electrostatic field, a current flow of these minority carriers crosses the junction. With an open circuit, a negative charge builds up in the n material and a positive charge in the p region. The resulting voltage is called the

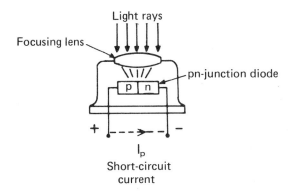

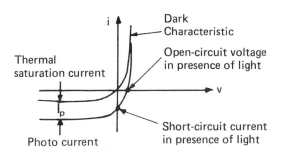

Figure 3-7. When light of the proper frequency band is allowed to shine on a pn junction, electron-hole pairs are generated in the transition, or depletion, region which produce a voltage across the junction that can cause current to flow in the external circuit.

photovoltaic voltage of the diode. Note that this voltage polarity forward biases the junction.

If the photodiode is then short circuited, this forward-bias voltage will cause the majority carriers in the rest of the semiconductor material that is distant from the junction region to flow and provide current to the external connection.

This is how solar cells for the generation of electric power from light work. Later sections of the book will take up discussions of solar cells and photodiode circuits and applications.

Metal-to-semiconductor junctions

Often rectification is desired in metal-to-semiconductor junctions, as in selenium and copper-oxide rectifiers, but there are many more instances in the design of solid-state devices where a metal-to-semiconductor junction should be ohmic (Figure 3-8). Until the theory of the solid state was better understood, the selection of materials to produce the desired effect—ohmic or diode action—was mostly a cut-and-dry process. Now, the exact junction behavior can be predicted by examining the energy level characteristics of the metal and semiconductor materials making up the junction.

In general, the average energy (Fermi level) of the electrons in a metal will not be the same as that of the electrons in a semiconductor at equilibrium (Figure 3-9(a)). Thus, when the metal and the semiconductor are placed into contact there is a transfer of charge across the contact which sets up an electrostatic potential and the other conditions associated with pn junctions, as previously described (Figure 3-9(b)). If the semiconductor is n-type material, the energy level diagram of the junction will attain equilibrium as shown and behave in nearly every way like a pn junction. The metal will assume the role of p material. The contact will rectify since it can "inject" holes into the n region when it is forward biased (metal positive and semiconductor negative) and it will extract holes from the n region when reverse biased.

Note that two new parameters have been introduced into the energy diagrams. They are ϕ_m and ϕ_s, the so-called thermionic work functions of the materials. The thermionic work function of a material is the minimum amount of energy that can cause the ejection of an electron from the surface of the material.

In Figure 3-8, $\phi_m > \phi_s$. If, on the other hand, $\phi_m < \phi_s$, then the

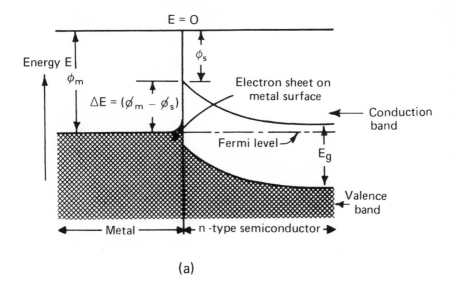

(a)

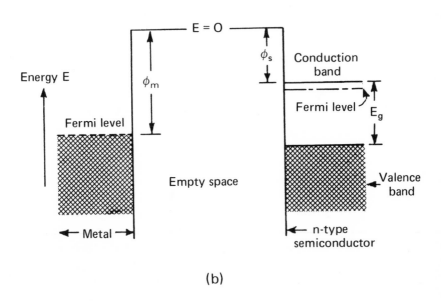

(b)

Figure 3-8. When a metal and an n-type specimen with thermionic work functions $\phi_m > \phi_s$ (a) are placed into contact (b), a junction is formed with characteristics very similar to a pn junction where the metal assumes a role similar to that of p material.

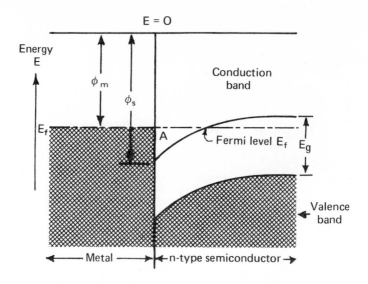

Figure 3-9. The junction of a metal and an n-type material, where $\phi_m < \phi_s$, allows free flow of conduction electrons between the metal and semiconductor. Energy level in region *A* fills up with electrons and the contact becomes ohmic.

energy diagram of Figure 3-9 would result and electrons would readily flow from metal to semiconductor in the process of establishing equilibrium. Thus a ready supply of conduction electrons is available for either direction of bias and the contact is ohmic.

We therefore see that a rectifying barrier *may* exist at an n-type semiconductor-to-metal contact if the thermionic work function ϕ_m of the metal is greater than the semiconductor work function ϕ_s. However, a rectifying contact cannot be formed by merely pressing together any ordinary specimens of materials. The usual result is an ohmic contact. At the very least, a good rectifying contact requires that all surface imperfections be polished or etched away and that the purity of the contacting materials be maintained over the whole area of contact. On the other hand, deliberate damage to the surfaces by sandblasting, welding or soldering will usually result in an ohmic contact.

A p-type material in contact with a metal will exhibit behavior very similar to that of n-type material, except that the criterion for obtaining an ohmic or rectifying junction is reversed. For a rectifying junction with p-type material, $\phi_s > \phi_m$.

Chapter 4

Creating Solid-State Devices from PN Junctions

Although the single pn junction is very useful and versatile in its own right, its potential for revolutionizing the electronics industry was realized only after it was combined with other junctions. Consider the reverse-biased junction in Figure 4-1(a). Only a small amount of current flows—the result mostly of the presence of minority charge carriers, as explained in Chapter 3. Add another junction to this diode structure, as in Figure 4-1(b), and provide it with a forward bias.

It would be natural to expect that a heavy current would flow in this forward-biased junction and that the reverse-biased section would continue with its small current. *This is completely incorrect.*

The unexpected happens when the center region is thin (usually less than 10^{-3} cm). For the shown pnp combination, the forward bias of the left-hand diode flows completely through both junctions. Only a small percentage of the current returns through the center region. With battery polarities reversed, an npn sandwich combination would behave similarly.

The left-hand region of the junction sandwich is called the emitter; the right-hand, the collector, and the center region is called the base. The combination is called a transistor. Current carriers that leave the emitter and pass into the base are considered as *injected* into the base. They become minority carriers in the oppositely doped material of the base. There is an average distance that such a current carrier will travel before it combines with the fixed ions of the crystal dopant and no longer acts as a current carrier. This distance is called the *diffusion* length. The length of time that a current carrier exists is called its *lifetime*.

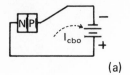

(a)

In a reverse-biased junction a small, so-called, saturation current flows because of minority carriers. To this, ordinary resistive leakage also contributes some current to constitute the total, I_{cbo}.

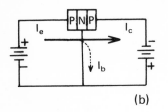

(b)

A forward-biased p layer added over the thin n layer injects current into original p material. Most of this forward-bias induced current passes right through to the reverse-biased section and only a fraction of this current flows in the n-material return path.

(c)   A pnp transistor and circuit symbol.

(d)   An npn transistor and circuit symbol.

Figure 4-1. A junction transistor consists of two p- or n-type materials that sandwich a very thin layer of oppositely doped material.

When the base region is much thinner than the average diffusion length, minority current carriers that are injected into the base can pass right through and be captured by the collector. This is the secret of transistor action. In a properly proportioned npn or pnp sandwich most of the carriers injected into the base (98 to over 99 percent) pass into the collector region.

To understand the ability of a minority current carrier to have a lifetime sufficiently long to pass through the base region, you must remember that the number of impurity atoms in both p- and n-type materials is extremely small compared to the number of intrinsic atoms in the crystal lattice, even in highly doped materials. A single free electron (majority carrier), wandering from an n region into p-type base material and thus becoming a minority carrier, will therefore on the average travel a considerable distance before it is captured to fill a vacant hole. Similarly, a hole in n-type material can wander among many lattice ions before it is likely to find and be filled by a free electron. It is thus evident that the greater the doping level, the shorter is the lifetime of such minority carriers. On a very thin path through the base region, most carriers can make it through to the collector region.

How transistors amplify

Most but not all of the minority of current carriers injected into the base reach the collector junction. Some few combine with majority carriers in the base, even if the base is very thin. The carriers that combine in the base produce an equivalent base-lead current, I_b. The current that flows in the collector is equal to the input, or emitter, current I_e minus the base current, or

$$I_c = I_e - I_b$$

Since I_c/I_e is less than unity, you might justifiably ask how a transistor amplifies. The trick is to control the base current.

Apply a dc signal to the base as in Figure 4-2. Let us say that the I_c/I_e ratio, which is called α (alpha), is 0.98. Since 98 percent of the emitter current goes into the collector, to increase the base current from an initial value of perhaps 0.2 mA to, say, 0.3 mA requires that the emitter current increases to 15 mA if its initial value is 10 mA.

Thus the ratio of the changes in I_c and I_b ($\Delta I_c/\Delta I_b$) equals 5/0.1. This is a current gain of 50. This ratio is called β (beta). These values of alpha and beta are for dc, since the bias and signal currents so far

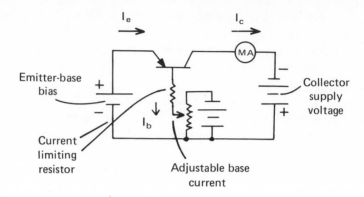

Figure 4-2. A small change in base current produces a large change in collector current.

have been dc. In practice, however, the alpha and beta terms more frequently refer to ac signals. Where the signal levels are small as compared to the bias levels, and the frequency is low, both ac and dc values for alpha and beta are nearly equal.

Power is amplified too

Because the emitter-base junction is forward biased, the resistance in this path is low as expected for a forward-biased diode. If an input signal is then fed to this junction, the input voltage $E_s = I_e R_e$. The output signal is then $E_o = I_c R_o$. Since $\alpha = I_c / I_e$, then

$$G_b = E_o / E_s = \alpha R_o / R_e$$

In a typical case, R_e might equal 100 Ω and R_o = 10 k Ω. The voltage gain would then be approximately 10,000/100 = 100, since $\alpha \approx 1$.

In addition to voltage gain, the difference in input and load resistance leads to the power gain:

$$G_p = \alpha^2 R_o / R_e$$

Transistor characteristics curves

Since the base-collector junction is reversed biased, a plot of collector current, I_c, versus collector-to-base voltage, E_c, looks like a reverse-biased diode. The current, after a small initial rise, remains almost constant and independent of voltage. This current is the equivalent of the saturation current of a diode (Figure 3-6). Like saturation current, the collector current is strongly temperature dependent, since the current is due mainly to intrinsic minority carriers.

If new current is introduced into the emitter junction, the thin base region allows most of these charge carriers to enter the collector region. The quantity of these additional current carriers that are injected from the emitter also is independent of the base-to-collector voltage and simply adds to the saturation current curve. Thus a family of I_c vs. E_c curves consists of a set of parallel plots—one for each value of emitter current, I_e (Figure 4-3).

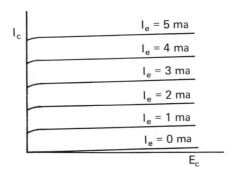

Figure 4-3. The family of common-base characteristics curves shows the relative independence of the collector current from changes in collector voltage.

Designing basic transistor circuits

There are three basic transistor circuits. Each is identified by the transistor element that is common to both the input and output. Very often this common element is connected to the circuit's ground. Figure 4-4 shows the fundamental configurations. The family of characteristics curves in Figure 4-3, however, applies to the common-base arrangement. Let's explore the properties of the common-base configuration first.

The best way is to analyze such an amplifier. Some simple graphical constructions on the characteristics curves can tell you how the

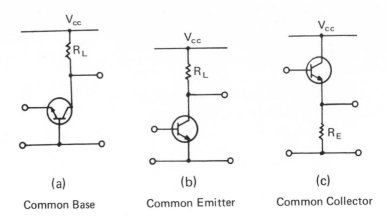

(a)   (b)   (c)

Common Base   Common Emitter   Common Collector

Figure 4-4. Basic transistor-amplifier configurations.

amplifier behaves—how to set the bias, how much collector current is drawn, how much power the collector must dissipate and how much gain the circuit will provide. Figure 4-5 is the circuit that we will study.

Start by drawing a load line on the characteristics curves. Since the supply voltage for the collector is 24 V, the line starts at the 24 V point on the E_c axis. The 6-k Ω load resistor determines the slope of the line and that it will intersect the I_c axis at $24/6000 = 4$ mA. All operating points lie on this load line.

Because transistors are primarily current devices, the quiescent operating point is established by means of a bias current. To obtain a maximum linear range of output voltage variation, the bias current into the emitter is selected somewhere in the middle of the load line, at about $I_e = 2$ mA. Thus the bias resistor is equal to $6/2 \times 10^{-3} = 3000$ Ω. The emitter resistance is usually very small (less than 50 Ω) and can be neglected.

If the transistor has an $\alpha = 0.98$, then I_c is about 2 percent less than I_e, or 1.96 mA. This produces a voltage drop across the load resistor of 9.8 V and leaves 10.2 V from collector to ground. The power dissipated at the collector is $P = IE = 1.96 \times 10^{-3} \times 10.2 = 19.9$ mW.

Note that the current gain is equal to α which is less than one. The common-base circuit is not an effective current amplifier. But the

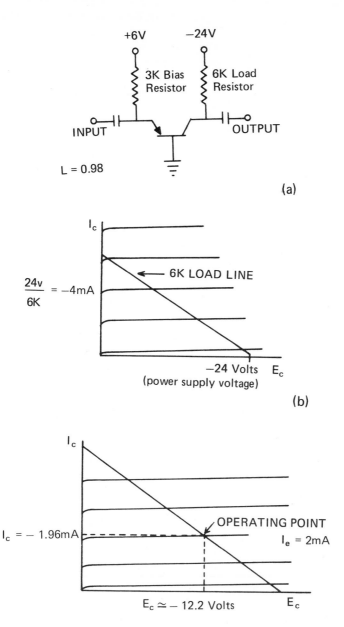

Figure 4-5. The transistor used in a common-base amplifier circuit (a) has the $I_c \cdot E_c$ characteristics curves (b) that are used to determine the essential parameters and operational capabilities of the circuit (c).

voltage gain can be quite large. It can be calculated from a knowledge of the input and output resistance. The output resistance is mostly the $6000-\Omega$ load resistance. The input resistance can be more precisely determined than the loosely stated less than 50 Ω previously given.

Basic semiconductor physics tells us that the average energy of a charge carrier in a conductor (or semiconductor) is

$$eV = KT.$$

K is a special number known as the Boltzman constant, e is the charge on an electron, T is the absolute temperature and V is the voltage drop across the conductor (see pages 00 to 00). Since in the emitter $V = I_e R_e$, then $R_e = (KT/e)/I_e$. At a room temperature of 25° C, the theoretical value for emitter resistance becomes approximately

$$R_e = 0.0256/I_e$$

Practical transistors will have emitter resistances that can vary appreciably from this value, but for first-order calculations, the values obtained with this equation are sufficient.

$$\text{Thus } R_e = 0.0256/2 \times 10^{-3} = 12.8 \ \Omega$$

The voltage gain can now be calculated:

$$G_v = \alpha \ R_0/R_e = 0.98 \times 5000/12.8 = 383$$

Common-base amplifier voltage gain value compares very favorably with that obtainable from vacuum tubes. Unfortunately, the low input resistance of 12.8 Ω is a severe limitation. A common-base amplifier must be fed from a source that is even of a lower resistance than this low input resistance to make effective use of the available voltage gain. This is often not practical; also, such signal sources are not common.

In spite of this, the common-base configuration finds applications in many circuits. One very important application is as a low-to-high impedance coupling circuit (Figure 4-6). Note that it provides gain without inverting the signal.

Common-emitters supply voltage and current gain

The common-emitter circuit is the more universally useful circuit. It has a reasonably high input impedance and provides both voltage and

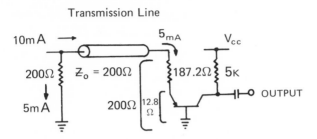

Figure 4-6. The common-base circuit can provide both low-to-high impedance coupling and gain. For example, a transmission line whose characteristic impedance is 200 Ω must be terminated with 200 Ω for proper performance. The common-base configuration easily accomodates this. If a signal source can supply a peak-to-peak 10 mA signal, and half of it flows in the transmission line's input and the other half into the amplifier at the output end, then a voltage gain

$$\frac{e_o}{e_i} = \frac{5 \text{ mA} \times 5000}{5 \text{ mA} \times 200} = 25$$

is obtained, besides the provision of the 200-Ω termination impedance.

current gain. In this configuration (Figure 4-4(b)) the signal is applied to the base and the emitter is usually returned to ground. Since the signal adds or subtracts from the very small base current, which is a fixed percentage of the collector current, variations of the base current cause proportionately much larger variations in the collector current. The configuration in Figure 4-2 and the associated explanation thus apply to the common-emitter circuit.

If now a family of curves is plotted for I_c versus E_c, but for selected base currents, I_b, instead of emitter current, it will be observed that the curves differ from the common-base curves in several ways (Figure 4-7).

- The flat parts of the curves have a steeper slope.
- The curves do not touch the I_c axis.
- The I_b parameters on the curves are much smaller currents than I_e.
- The collector current for $I_b = 0$ is much larger than for $I_e = 0$.

The importance of these differences is best explained with practical examples.

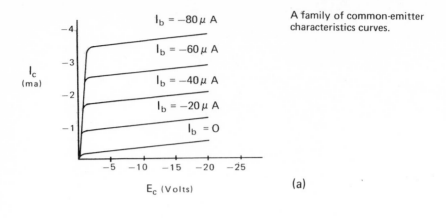

A family of common-emitter
characteristics curves.

(a)

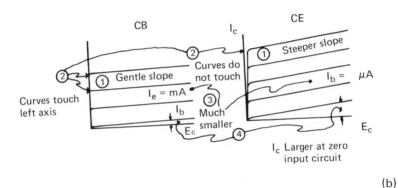

(b)

Figure 4-7. The characteristics curves for the common-base (CB) and common-emitter (CE) circuits differ in many important ways.

Analyzing a common-emitter amplifier

As in the case of the common-base circuit, the common-emitter circuit (Figure 4-8) can be best analyzed with some graphical constructions on the common-emitter characteristics curves. Again begin by drawing the load line on the family of curves, and select the quiescent operating point somewhere midpoint along the load line for maximum linear E_c - I_c swing range.

Note that the selected bias current is 40 μA, the collector current is 2 mA and the collector voltage is 10 V. The power dissipated at the collector is: $P = EI = 10 \times 2 \times 10^{-3} = 20$ mW. The bias current is

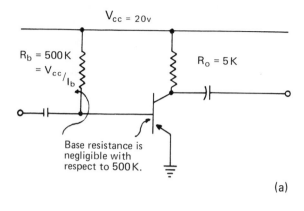

(a)

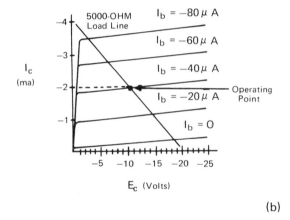

(b)

Figure 4-8. The family of characteristics curves for the common-emitter circuit provides the data needed to calculate common-emitter performance.

determined by $R_b = 20/40 \times 10^{-6} = 0.5$ M Ω; the base resistance can be neglected in this calculation.

Earlier it was stated that the emitter current divides itself into a collector-current flow and a smaller base-current flow. The collector current is αI_e and the base current is $I_e - \alpha I_e = I_e (1 - \alpha)$. Thus $I_c/I_b = \alpha I_e/I_e (1 - \alpha) = \alpha/1 - \alpha = \beta$. It can thus be seen that as α approaches unity, the value of β can rise quite high—200 to 300 is not unusual. To obtain an actual common-emitter current gain equal to β, the load resistance must be very low or shorted out. A vertical load line would represent this condition.

Since the load line is not vertical—it has a slope that corresponds to 5000 Ω —the actual current gain is somewhat less and can be determined from the family of curves.

From the curves we see that, for a signal input that has a peak-to-peak swing of 40 μA, the output swing is 1.6 mA peak-to-peak. The actual current gain is then $G_c = 1.6 \times 10^{-3}/40 \times 10^{-6} = 40$, rather than the $\beta = 50$ value.

The voltage gain can be calculated from the following equation:

$$G_v = I_c R_o/I_b R_b = G_c R_o/R_b$$

where $G_c = I_c/I_b$.

Note that R_b, the resistance when looking into the base, has replaced R_e, and G_c has replaced the α that was formerly used to calculate the voltage gain of the common-base amplifier.

The base input resistance can be determined as follows:

$$R_b = E_{be}/I_b = E_{be}/I_e (1 - \alpha) = R_e/(1 - \alpha)$$

V_{be} is the base-to-emitter voltage and R_e can be the theoretically determined value (12.8 Ω) established previously. Another way of writing the equation for the base resistance is

$$R_b = R_e (1 + \beta),$$

since $1 + \beta = \alpha/(1 - \alpha) + 1 = [\alpha/(1 - \alpha)] + [(1 - \alpha)/(1 - \alpha)] = 1/(1 - \alpha)$.

Then the voltage gain for a common-emitter amplifier can be written as

$$G_v = G_c R_o/R_e(1 + \beta)$$

For the amplifier under examination

$$G_v = 40 \times 5000/12.8(50 + 1) = 306$$

Besides having both current and voltage gain, another very important advantage of the common-emitter configuration is that the base input resistance is $R_e (1 + \beta)$. Where the common-base circuit of the previous example had an input resistance of only 12.8Ω, the common-emitter arrangement is 51 times as great, or 652.8Ω. Higher betas will naturally provide higher input resistances, which can become several thousand ohms. With a resistor placed in series between the emitter and ground to make R_e look bigger, the input resistance can approach that of a vacuum tube, but at the expense of gain. More will be said about this when we examine the common-collector circuit.

The common-emitter circuit that we have thus far studied in Figure 4-8 is not a very practical circuit, however.

Making the common-emitter circuit practical

The biasing method used in Figure 4-8 is an extremely poor way to bias a transistor. Its sole virtue was that it allowed simple calculations and an easy explanation of how the basics of the circuit work. Such a biasing method would be very sensitive to changes in temperature. Furthermore, practical transistors have rather wide tolerances. The published characteristics curves are at best only an average. The particular transistor that you may have probably does not conform closely to the published curves. Practical circuits must be able to tolerate wide transistor tolerances and not be sensitive to transistor replacement; also, the circuit should tolerate wide temperature variations. A proper bias arrangement goes a long way towards the achievement of both these goals.

The 0.5 MΩ resistor used to bias the circuit of Figure 4-8 results in an amplifier that may work well at room temperature, but at, say, 50°C, it would become completely inoperative; or it may work well with one carefully selected transistor but not with another, even though it was chosen at random from the same batch.

For a practical bias arrangement on a common-emitter configured amplifier, consider the circuit in Figure 4-9(a). Note that the bias voltage divider delivers 2.4 V at the transistor's base, if we neglect the small current into base—$I_b = I_c/\beta \approx 2.5$ mA/50 ≈ 50 μA. The divider resistances carry about 0.5 mA while the base of the transistor draws only microamperes.

If the base voltage is 2.4 V, the emitter is nearly 2.4 V, since the emitter-base junction is forward biased. The voltage drop from emitter to base is very small—less than 0.2 V for most transistors. This is a negligible amount for the purposes of this approximate calculation.

The 1000-Ω resistor to the emitter will then carry about 2.4 mA to produce this 2.4-V potential drop. Since I_c almost equals I_b—$\alpha = 0.98$—let's assume that the collector current is also 2.5 mA. The voltage drop across the load resistor is then 12.5 V.

To compute the gain of the amplifier, redraw the circuit as in Figure 4-9(b). Since the input signal is an ac signal and any power supply suitable for amplifiers must be well bypassed to ground for ac signals, the bias voltage divider appears to the signal as if it were a

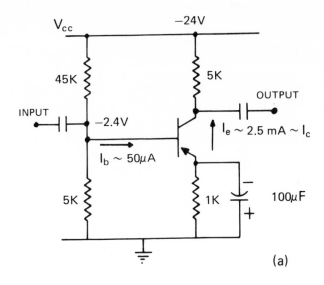

(a)

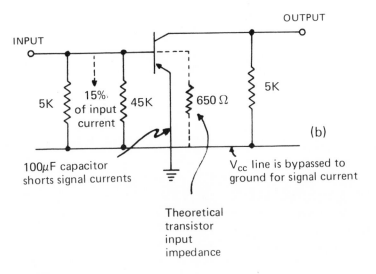

(b)

100μF capacitor
shorts signal currents

V_cc line is bypassed to
ground for signal current

Theoretical
transistor
input
impedance

Figure 4-9. A common-emitter amplifier can be stabilized and made less sensitive to temperature and transistor selection by the practical biasing arrangement shown in (a). The 1-k Ω resistor in the emitter path makes the dc behavior of the circuit fairly independent of the transistor's dc variability, and the bias voltage divider maintains the operating point much more stably than does a single series resistor as used in Figure 4-8(a).

parallel circuit. The load resistor is also effectively returned to ground. And the large capacitor across the emitter resistor bypasses this 1000-Ω resistor for ac signals.

If the previously calculated base input resistance is rounded off to 650 Ω, then roughly 15 percent of the input signal is lost into the bias resistors. The current gain of 40 that was computed for the simplified amplifier is thus reduced to about 34 for the practical circuit. This is the sacrifice that must be made to attain more stable performance.

The reduced current gain also results in a reduced voltage gain, since $G_V = G_C R_O / R_e (1 + \beta)$, and G_C is now less. Substitute $G_C = 34$ into the equation and $G_V = 260$.

The common-collector has high input resistance

The common-base amplifier has large voltage gain but less-than-unity current gain and low input resistance. The common-emitter configuration can provide both voltage and current gain and has moderate input resistance. The *common-collector* fills in the remaining possibilities—*it has high current gain, less-than-unity voltage gain* and *high input resistance* (Figure 4-10). It is often referred to as an *emitter-follower* circuit because its characteristics resemble the vacuum-tube cathode follower circuit.

The name emitter-follower is very appropriate since, as we al-

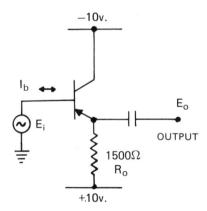

Figure 4-10. The common-collector, often called the emitter-follower circuit, provides a high input impedance, low output resistance and slightly less than unity voltage gain. It is used as an input buffer for other high-gain circuits that have lower input impedances.

ready know, the emitter and base voltages are nearly equal because of the forward-biased junction they form. A signal on the base will cause the emitter to closely *follow* its variations—$E_o \sim E_i$. Since $I_b = I_e (1 - \alpha)$ and $I_e = E_o/R_o$, we can write $I_b = E_o (1 - \alpha)/R_o$, or substituting E_i for E_o

$$E_i/I_b = R_i \sim R_o/(1 - \alpha) \sim R_o (1 + \beta)$$

With a load resistor, R_o, of 1500Ω, the effective input resistance, R_i, of the common collector circuit becomes $1500 \times 51 = 76,500 \ \Omega$.

The input voltage, E_i, is slightly larger than the output voltage by the voltage drop in the base-to-emitter junction, E_{be}. Thus

$$E_i = E_o + E_{be}$$
$$\sim E_o + E_i (1 - \alpha)$$
$$\alpha \sim E_o/E_i \sim G_v \sim 0.98,$$

since $E_{be} = I_b R_b = I_e (1 - \alpha)R_b = E_o(1 - \alpha) \sim E_i (1 - \alpha)$. These approximations assume that the input signal source has negligible internal resistance.

Field-excited transistors (FETs)

Transistors thus far discussed depend on the flow of both majority and minority charge carriers for their operation and are thus called *bipolar*. An often used acronym for the common transistor is BJT—*bipolar junction transistor*. A different physical arrangement of p and n material (Figure 4-11(a)) results in a *unipolar* transistor which is more commonly called an FET—field-excited transistor—or more specifically, a *JFET*. The J stands for junction. A second type of FET is designated *IGFET* for *insulated-gate field-effect transistor*, which insulates the gate from the source-drain channel.

The JFET consists of a thin bar of n- or p-type material. Ohmic contacts to opposite ends of this bar, which are called the *source* and *drain*, are analogous to the emitter and collector in a BJT. In the center of the bar, on opposing surfaces, oppositely doped semiconductor pads form a *gate* (analogous to the base in a BJT). If an n-type bar is used, the gate pads are made of p-type material, and vice versa. In the example of Figure 4-11, because the channel between the two gate pads is made of n-type material, the transistor is an n-channel JFET.

In normal operation, the drain of an n-channel JFET is connected to the positive polarity of a power supply and the gate is held negative

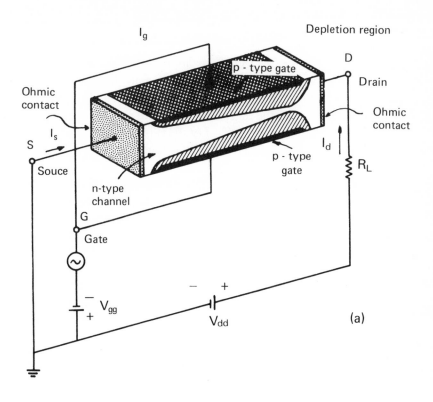

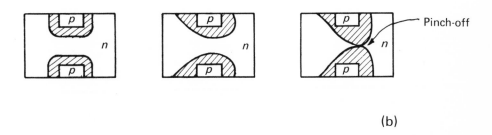

Figure 4-11. An n-channel, field-excited transistor's basic structure (a) provides a bilateral resistive path between the ohmic contacts, source and drain. As gate bias, V_{gg}, is increased, the depletion region becomes larger until current in the channel is pinched off (b).

with respect to the source. These polarities are, of course, reversed for a p-channel JFET.

Thus, since the p-material pads that form the gate have a negative bias, the pn junctions to the bar are reverse biased.

It is interesting to compare this biasing to that for normal amplifier operation of a BJT. The BJT is forward biased. This fundamental biasing difference is the reason that the input resistance of an FET is very high compared to the BJT. The BJT is thus a low-impedance, current-operated device, while the FET is high-impedance and voltage operated.

Another basic difference is that the source-drain channel is ohmic, or bilateral, (conducts in either direction), whereas the emitter-collector path is unilateral, or diode acting. The source-to-drain current path passes through only one type of semiconductor material; therefore only the material's ohmic properties are involved in the conduction of current.

However, the current must pass between the closely spaced gate pads. Since the pads are reverse biased, almost no current flows into them, but an electric field bridges the junction to the bar material. As in the case of a reverse-biased diode, a *depletion* region is created in the channel that is characterized by the creation of uncovered ions and reduction of free carriers. In other words, the channel's conductivity is reduced.

The greater the gate reverse bias is, the greater is the region of depletion and the higher the resistance in the source-to-drain path. When the depletion regions that extend from each gate pad touch, practically no current flows in the channel. This gate bias voltage is called the pinch-off voltage, V_p.

Note in Figure 4-11(b) that the current channel between the depletion areas narrows near the drain end of the channel. This happens because of the IR drop of voltage along the channel so that near the source end there is less field strength to the gate pads and thus a thinner depletion region. For a particular reverse-bias gate voltage, low drain currents provide small IR drops and have little effect on the thickness of the depletion region. The drain current therefore initially increases proportionately and steeply with increasing drain voltage (Figure 4-12), but as the drain current continues to rise, this IR voltage drop starts to affect the depletion configuration, as shown. This "pinching" effect further limits drain-current rise; the drain current vs. voltage curve flattens out to become almost independent of the drain voltage.

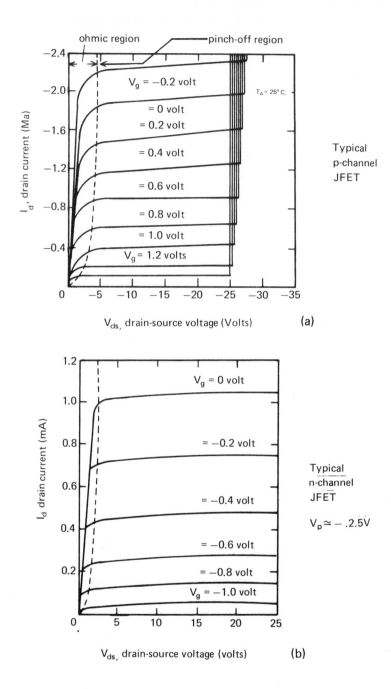

Figure 4-12. Depletion JFETs may be operated with a few tenths of a volt forward gate bias (a). The steeply rising left-hand portion of the curves are called the ohmic region and the flat right-hand portion, the pinch-off region.

This flat part of a drain characteristic curve is called the pinch-off operating region. The pinch-off region is where the drain voltage, V_d, is greater or equal to the pinch-off voltage V_p less the gate voltage, V_g, as follows:

$$|V_d| \leq |V_p| - |V_g|$$
(only magnitude, not polarity, is considered)

The steeply rising portion of the curve is called the ohmic region. Figure 4-12 shows the pinch-off and ohmic regions separated by a dotted line which connects the centers of the "knees" of each I_d-V_d characteristic curve. In the ohmic region, the I_d-V_d relationships have slopes that range from $(100 \ \Omega)^{-1}$ to $(100 \ k\Omega)^{-1}$, and the slopes pass right through the origin since the source-drain current path conducts bilaterally.

It is interesting to note from the characteristics curves in Figure 4-12 that forward-bias operation is also possible.

Not only does the JFET present a very high input resistance for reverse-bias operation, but also for small forward-bias voltages when the voltage is less than a few tenths of a volt. This is particularly true of silicon JFETs, since at these small voltages a forward-biased silicon diode remains almost an open circuit at room temperatures.

Insulating the gate in an FET

Rearrange the p and n material as in Figure 4-12 and provide a very thin insulating material under a single gate pad and you now have an IGFET—insulated-gate field-effect transistor. Now the drain and source contacts to the current channel region are via pn junctions at both places, whereas the JFET used ohmic contacts. The gate pad material is metal and the insulation is the metal oxide, silicon dioxide; hence the often used designation MOSFET—metal-oxide-semiconductor field-effect transistor.

The source and drain semiconductor regions are heavily doped semiconductor regions of an opposite conductivity type from the substrate material. The source and drain paths run through back-to-back diodes and the drain current is practically zero when the gate voltage is zero. In the Figure 4-13 configuration, when the gate-to-substrate voltage is made sufficiently large, the region between the source and drain semiconductor materials is enhanced and attracts free holes to form a p-conduction channel. Since conduction in the source and drain p

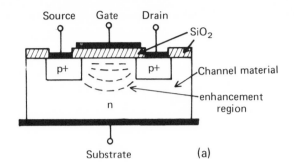

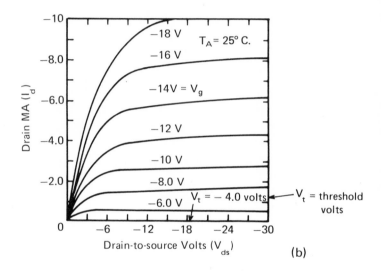

Figure 4-13. An IGFET's basic construction (a) uses SiO₂ insulating material between the gate pad and the channel semiconductor material. When the gate voltage rises above a threshold level, the channel between source and drain is enhanced with current carriers. Note that the gate and drain voltages are of the same polarity and order of magnitude (b).

material is also via holes, diode action ceases and the conduction channel becomes essentially ohmic to current flow—thus the designation, enhancement type MOSFET.

The gate voltage needed to form the channel is called the threshold voltage, V_t. An increase in the negative gate voltage above this minimum threshold voltage further enhances the presence of free holes to increase the flow of drain current.

MOSFETs may be either p- or n-channel devices. Except that gate and drain polarities are negative for p-channel MOSFETs and positive for n-channel types, modern units have comparable characteristics. Many types are made in pairs that complement each other (n- and p-channel units of similar characteristic) and their opposite voltage requirements permit some very useful circuit combinations.

What is enhanceable can also be depleted

If the source, drain and channel regions are made of similar materials, the MOSFETs become a depletion type. Similar to JFETs, they conduct substantial current with zero gate bias, but unlike the JFET, the MOSFET has an insulated gate. The cross section of an n-channel depletion is shown in Figure 4-14. The source and drain regions are more heavily doped than the channel material, and the n material lies in a thin layer atop p material. A negative gate voltage can drive free conduction electrons from the n-channel and thus can deplete the region to reduce its conductivity. When the gate voltage becomes sufficiently negative, conduction is pinched off as in the JFET. As before, this voltage is called the pinch-off voltage V_p. The manufacturer usually will provide a rated pinch-off voltage at a specified small drain current and drain voltage.

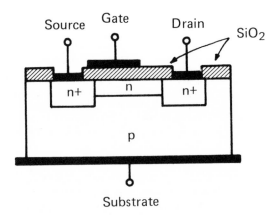

Figure 4-14. An insulated-gate depletion MOSFET uses the same type of material for the source, drain and channel. However, the regions adjacent to the source and drain are more heavily doped than the channel.

Because of the insulated gate, the depletion MOSFET, unlike the JFET, can operate with both positive and negative gate voltages without the fear of drawing gate current (Figure 4-15). Both MOSFET and JFET drain characteristics curves have similar shapes for about the same reasons.

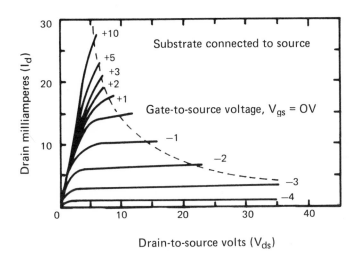

Drain-to-source volts (V_{ds})

Figure 4-15. MOSFETs of the depletion type may use either positive or negative gate voltages, and the insulated gate draws practically no current.

Although in theory a p-channel depletion-type MOSFET is also possible, practice has not been able to produce a p-type channel on an n-type surface properly. Substrate *surfaces* tend to become n-type no matter what the substrate material is.

Problems with MOSFETs

Though the MOSFET's extremely high input impedance is the source of its main advantage over bipolar transistors, it is also the cause for a high rate of damage to the MOSFET when the device is improperly handled. Static charges that accumulate on plastic wrapping or containers, slight leakages in soldering irons from the power line and static accumulations on the technician or wireman as he walks across the floor are enough to puncture and destroy the input gates of

MOSFETs; therefore, their leads should be shorted together at all times other than when under test or in actual use.

The units should not be handled unnecessarily and should only be picked up by the can—not by the leads. Soldering irons and all tools that come in contact with MOSFET leads should be well grounded. Good practice would prescribe a grounding bracelet for the wireman.

Because of possible damage by voltage transients, MOSFETs should not be inserted or removed from circuits with the power on. This is probably a good rule to follow even for bipolar transistors.

When compared to JFETs, MOSFETs are found to generate somewhat more noise in high-gain amplifiers. While depletion MOSFETs are more easily damaged by nuclear radiation than JFETs, enhancement MOSFETs are more resistant to such damage than JFETs.

The UJT—a transistor that switches

BJTs and FETs may be used in switching circuits—applications that require only two states, ON and OFF, or high and low impedance. But they are primarily classified as linear control devices. The unijunction transistor (UJT), however, is a bistable device that is specifically intended for switching action.

The physical arrangement of the junctions in a UJT is very similar to that of a JFET. Figure 4-16(a) shows the construction used in most commercial UJT devices. A high-resistivity bar of n-type material, usually of silicon (typically $5 \times 10 \times 50$ mils) and called the *base*, has a p-type *emitter* inserted somewhere near its middle. Ohmic contacts at the opposite ends of the bar are called *base-1* and *base-2*. The emitter-junction region is much smaller in the UJT than in the JFET, and the UJT emitter is forward biased during operation whereas the JFET gate is normally reverse biased.

As symbolically shown in Figure 4-16(b), the current path between base-1 and base-2 is ohmic, and the path between the emitter-junction point e and base-1 acts as a variable resistor that is dependent upon the emitter current. The diode between the input terminal and the emitter-junction point e represents the emitter-base pn junction.

The silicon bar between the bases acts as a voltage divider to the usually positive voltage on base-2. When the emitter voltage is less than the voltage at point e, the emitter diode is reverse biased and very little current flows in the emitter circuit. If the silicon bar divides the

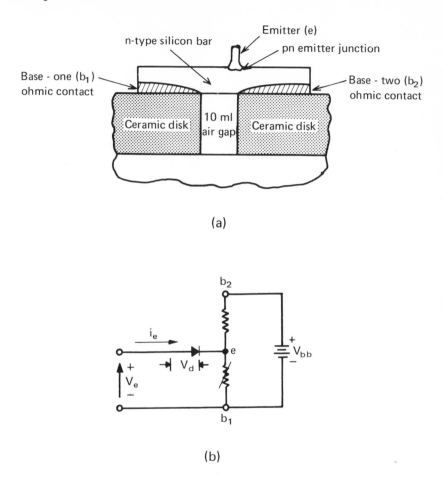

Figure 4-16. In a unijunction transistor, the physical structure (a) resembles a JFET, and the current path between base-2 and base-1 is ohmic (b) also as in a JFET.

voltage at point e in a ratio equal to $n\mu$, called intrinsic stand-off ratio, then when the emitter voltage V_e is slightly greater than $n\mu V_{bb}$ the emitter diode is forward biased and an appreciable current flows through the emitter diode to base-2.

Hole current carriers are injected into the bar from the p material of the emitter and electrons flow into the bar from base-1 to neutralize them. This increase in current carriers in the base-1-to-emitter region causes a reduction of this region's resistance so that the voltage re-

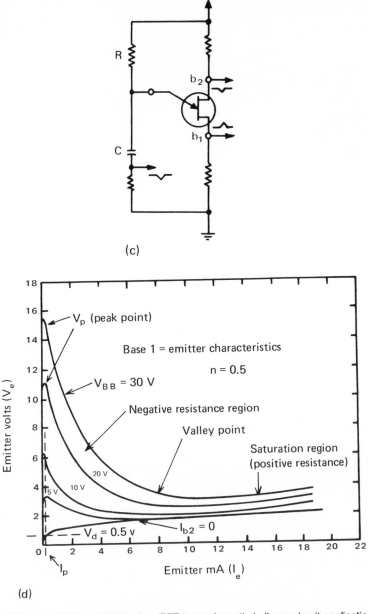

(c)

(d)

Figure 4-16 (cont.). While the JFET is used mostly in linear circuit applications, the unijunction is a triggerable device for switching applications as in this relaxation oscillator (c), which makes use of the unijunction's negative-resistance characteristic (d).

quired to forward bias the diode becomes less. In this way, an emitter-current rapidly increases as the emitter-voltage decreases. This voltage-current behavior is called a negative resistance.

In a typical commercial unijunction transistor, the base-1-to-emitter resistance might be about 5000 Ω when the emitter current is near zero. This resistance sharply drops to the neighborhood of 50 Ω when the emitter voltage rises sufficiently to forward bias the emitter diode. Thus a capacitor connected between the emitter and base-1, if charged through a resistor from V_{bb}, would be very rapidly discharged when the voltage across it reaches the forward-bias point (Figure 4-16(c)), V_p, where

$$V_p = nV_{bb} + V_d$$

V_d is the voltage drop across the emitter-to-bar junction at current I_p (Figure 4-16(d)). V_d can be measured when I_{b2} (current in base-2) is zero.

The SCR is a multi-junction transistor

Stack four semiconductor layers in series—pnpn—and you have made a *four-layer*, or *Shockley*, diode. Add a terminal (gate 1) to the second p layer, and you have created a gate-controlled, reverse-blocking *thyristor*, or *silicon controlled rectifier* (SCR) as it is more frequently called, since these devices are usually fabricated from silicon and widely used as rectifiers (Figure 4-17(a)).

The four-layer diode is analogous to a gas discharge tube, like a neon lamp, which conducts (fires) above a specific threshold voltage. The SCR is similar in behavior to a thyratron, where the gate controls the firing voltage.

The easiest way to understand how four-layer devices work is to symbolically separate the material of the layers into two three-layer transistor devices as in Figure 4-17(b). Figure 4-17(c) is a circuit diagram that is equivalent to the two-transistor representation, and shows the two transistors in a positive feedback connection. Although a second gate 2 is shown, let's ignore it for the present. If a_1 and a_2 are the current gains of the two transistor sections, with each gain value less than unity, then when

$$I_g = a_1 a_2 I_g + I_o,$$

where I_g is the current into gate 1, the circuit becomes self-feeding and

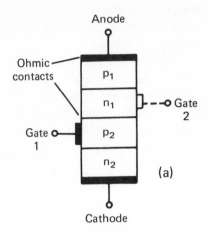

(a)

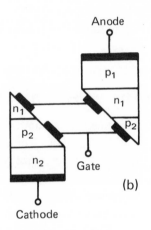

(b)

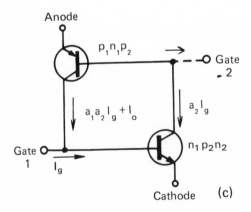

(c)

Figure 4-17. An SCR, often called a thyristor, is a four-layer device (a) that can be represented by an equivalent configuration (b), which operates as two transistors in a

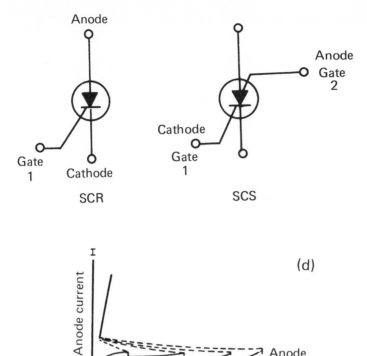

Anode

Anode
Gate
2

Cathode

Gate
1

Gate
1   Cathode

SCR   SCS

(d)

$I_g = 100\mu A$ $I_g = 0$

Thyristor breakover as function of gate current

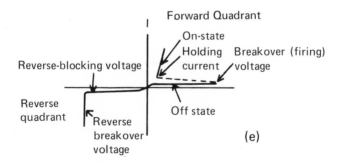

Forward Quadrant

On-state

Holding   Breakover (firing)

Reverse-blocking voltage   current voltage

Reverse
quadrant   Off state

Reverse
breakover   (e)
voltage

positive feedback circuit (c). The characteristics of a typical SCR's anode/cathode voltage
and current are shown in (d) and (e).

suddenly fires, or turns ON. Otherwise the $n_1 p_2 n_2$ section is cut off and the anode to cathode path does not conduct. Only a small leakage current, I_O, flows. After firing, the device can conduct a heavy current—in large sized SCRs, as much as 100 A—between its anode and cathode.

If we solve for I_g, we obtain

$$I_g = I_O/(1 - a_1 a_2)$$

Note that when the product $a_1 a_2$ is close to unity, this makes the denominator close to zero. For a small leakage current, I_O, the gate current can be extremely small (tens of microvolts) to fire the device.

Some four-layer devices have two gates, as also shown in Figure 4-17. Devices with this arrangement are called *silicon controlled switches* (SCS), they provide increased circuit-design flexibility. Gate 1 is called the *cathode gate* and gate 2, the anode gate. Where gate 1 fires the SCS with a positive pulse, gate 2 requires a negative pulse.

In a Shockley diode, where there are no gate connections, as the applied anode voltage is increased, I_O increases slowly. When the equality of the first equation is established, the diode fires at an anode voltage called forward-breakdown voltage. Typical pnpn diodes have ON voltages in the order of 1 V. SCRs can also be fired this way, but they often easily block as much as 1000 V.

Because of low impedances in the ON condition, both diodes and SCRs must be operated with a series resistance that is large enough to limit the anode-to-cathode current to a safe value.

Once fired, SCR and Shockley devices stay on until the anode current is decreased below the minimum holding current. Commercial devices normally require the order of several milliamperes to hold ON.

SCS units, however, can also turn OFF via their gates. Either a positive pulse to anode-gate 2 or a negative pulse to cathode-gate 1, which are opposite in polarity to turn-ON pulses, turns an SCS OFF.

Rate effects can cause problems

A rapidly varying anode voltage can cause a four-layer device to turn on, even though the voltage level never exceeds the forward breakdown voltage. Because of capacitance between layers, a current large enough to cause firing can be generated in the gated layer. Current through a capacitor is directly proportional to the applied voltage's rate of change, designated as dv/dt in differential calculus notation.

Turn-on by the dv/dt route can be accomplished with as little as a few volts per microsecond. Since a microsecond is a very small number, when it is divided into even a few volts, the rate of change becomes millions of volts per second—a very large number. Thus the dv/dt rating of a four-layer diode or SCR is very important and must be taken into account in circuit designs.

Another rate effect that is very important and also must be carefully considered is the anode-current di/dt rating. This rating is particularly important in circuits that have low inductance in the anode-cathode path, since then there is danger of excessive di/dt because inductance is needed to limit the rate of current rise when the device fires.

When an SCR is fired, the region near the gate conducts first and then the current spreads to the rest of the semiconductor material of the gate-controlled layer over a period of time. If the current flow through the device increases too rapidly during this period of time there will be too high a concentration of current near the gate and possible damage because of localized overheating. Gates designed to handle currents higher than that required to only fire the SCR will enable the device to withstand higher di/dt because the initial current area is then larger.

Turn-off takes time

As may be understood from the two-transistor representation of four-layer devices, once they are fired and become self-feeding in gate current, removal or even reverse biasing of the gate will not turn off the device. This is analogous to the behavior of thyratrons. Turn-off requires the reduction of the anode current below the holding current. When the device is used with an ac voltage on the anode, the unit turns off on the negative half of the voltage cycle. In dc switching circuits, other means must be used to reduce or reverse the anode voltage for turn-off.

Two important timing specifications apply to SCR turn-off characteristics: first, turn-off time, which is the minimum time between the points when the forward anode current becomes zero and when the device will block reapplied forward voltage without turning ON. This time is measured in tens of microseconds; second, reverse-recovery time is the time after forward conduction ceases, when the device can block reverse voltage in a rectifier circuit. Reverse recovery is typically in the order of several microseconds.

Switching bilaterally

The SCR can control only one-directional or dc current, but there are many applications that require the control of ac source power. This can of course be done by the use of two SCRs in parallel (back to back) with one device conducting one way and the other oppositely arranged (Figure 4-18(a)).

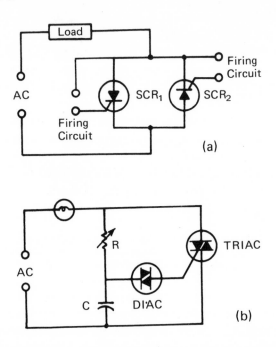

Figure 4-18. For control of ac power, two SCRs may be operated back to back (a). Diacs and triacs can provide similar bilateral ac performance in single packages (b).

By judicious arrangement of p- and n-type materials and junctions, however, a single bidirectional device can be built. Figure 4-19 shows a bilaterally conductive schematic arrangement that behaves very much like two four-layer diodes (diacs) or two SCRs (triacs) in parallel and oppositely conductive (Figure 4-18(b)). When terminal A is positive and above the breakover voltage, a path through $p_1 n_1 p_2 n_2$ conducts. When terminal B is positive, path $p_2 n_1 p_1 n_3$ can conduct. Each path is the equivalent of a four-layer diode as previously described (Figure 4-19).

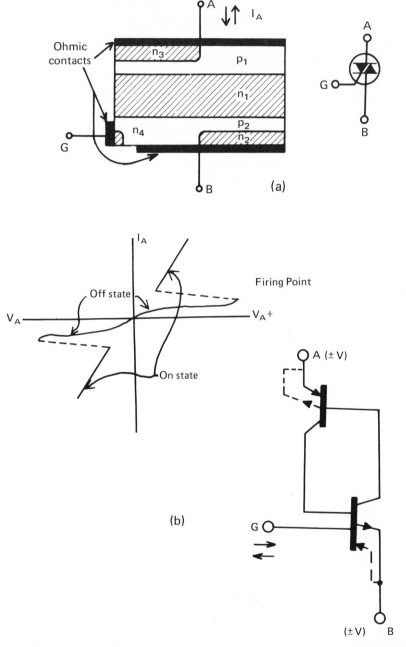

Figure 4-19. A combination of n and p layers arranged as in (a) provide bilateral thyristor performance (b).

A third element, terminal G, when it is sufficiently positive, will enable the $p_1n_1p_2n_2$ path to fire at a lower voltage than when it is zero. This action is almost identical with that of the SCR shown in Figure 4-17.

When terminal G is made negative and terminal B is positive, the firing point is lowered. This creates a bilateral control device in a single package.

The polarity relationships just described are the preferred gate-power polarities. The triac can operate with the opposite of these preferred polarity relationships, but then the required gate currents for proper performance are larger.

Most home lamp dimmers and solid-state small-motor speed controls use the triac. Figure 4-18(b) is a simple practical circuit. The variable resistor, R, and the capacitor, C, form a phase-shift network which can control the point in the applied sine wave at which the triac fires. The later it fires, the lower is the average current that flows and the dimmer is the lamp in the circuit.

Chapter 5

Hints for Designing with Transistors

Not all transistor circuits operate at a constant room temperature, at low audio frequencies and with signal levels that are small compared to the bias voltages. Practical circuits must deal effectively with wide temperature swings, high frequencies and large signal levels. But before circuits can be designed to use a transistor's full capabilities, the transistor's behavior under these conditions must first be clearly understood.

How temperature affects transistors

In Chapter 4, transistors were shown to be made of a combination of pn junctions similar to those used in diodes. It was pointed out that a reverse-biased junction normally passed a small current. This reverse current, I_{cbo}, has two parts. First, ordinary resistive leakage provides some of the current (Figure 5-1(a)). This current varies with voltage in accordance with Ohm's law. The second current component, called *saturation current*, results from the presence of minority carriers in the semiconductor material (Figure 5-1(b)). As previously explained, the quantity of minority carriers is highly dependent upon temperature but rather independent of applied voltage, so that saturation-current depends mainly upon temperature (Figure 5-1(c)).

Since in a transistor the base-collector junction is normally reverse biased, a transistor with its base grounded will behave similarly to a reverse-biased diode (Figure 5-2(a)). Note that the emitter is open circuited, so that no current passes through the emitter-base junction.

Most often a transistor is operated with the emitter grounded

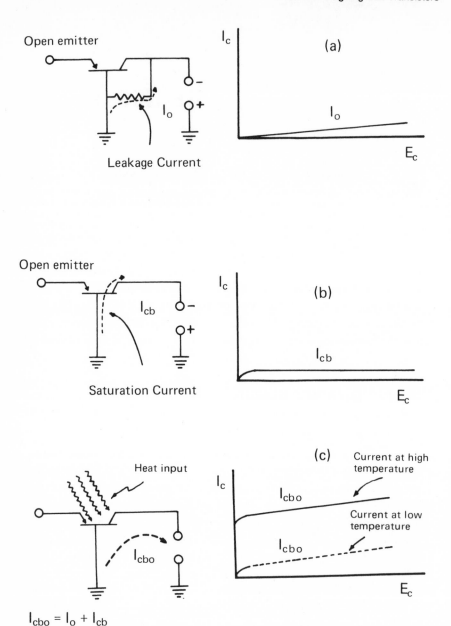

Figure 5-1. The reverse-biased collector with zero emitter current has a minimum current flow that depends upon ohmic leakage (a) and the reverse saturation current (b) of a reverse-biased diode. Saturation current is very sensitive to temperature variations (c).

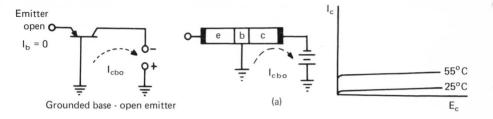

Emitter open $I_b = 0$

I_{cbo}

Grounded base - open emitter

I_{cbo}

(a)

55°C
25°C

E_c

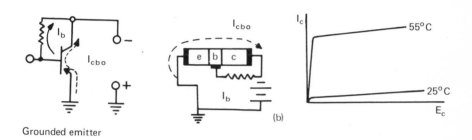

Grounded emitter

I_{cbo}

I_{cbo}

I_b

(b)

55°C

25°C

E_c

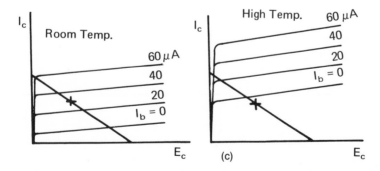

Room Temp.

60 μA
40
20
$I_b = 0$

E_c

High Temp.   60 μA
40
20
$I_b = 0$

(c)

E_c

Figure 5-2. When the emitter (b) is grounded instead of the base (a), the circuit is much more sensitive to temperature. A family of curves for an emitter-ground circuit (c) is displaced upward by an elevated temperature. This displacement can seriously disturb the performance of an amplifier circuit.

(Figure 5-2(b)). In this arrangement, however, the reverse current flows through the base-emitter junction as well as the base-collector junction. As we have learned, base-emitter current is amplified by β in the collector of a transistor.

Though an unstabilized base-grounded circuit is sensitive to temperature, the more widely used emitter-grounded circuit is even more sensitive. In almost every practical circuit, the design must include some means for temperature stabilization.

Older germanium transistors were highly sensitive to temperature. They exhibited large changes in I_{cbo} current. Newer germanium versions are much improved. Modern silicon transistors have much smaller leakage currents (nA vs μA) and present considerably fewer design problems because of I_{cbo} changes.

Dealing effectively with temperature

A good rule of thumb to remember is that for the unstabilized base-grounded circuit, as in Figure 5-1(a), the collector current doubles for about every 10°C increase in junction temperature; therefore, between 25°C and 55°C—a modest 30° rise—the current increases eight times.

For the emitter-grounded circuit, Figure 5-2(b), this increase is further elevated by the effects of β. Obviously, if uncontrolled, such behavior can seriously disrupt the functioning of any circuit and worse, the transistor can be destroyed because increased heating caused by increased current can cause a further increase in current—and so on.

To control transistor heating, we must first understand how to determine how much heat a particular transistor can safely withstand. This information is usually listed in the transistor manufacturer's data sheet as a *maximum allowable operating junction temperature* and a factor called the *thermal resistance* of the junction to the transistor case or to the ambient air (usually still air).

When a transistor is not drawing current, the junction temperature and the case and ambient temperature eventually become equal; however, when power feeds into the transistor, heat dissipated at the collector junction raises its temperature. The amount of rise is calculated with the use of the thermal resistance factor. A typical small-signal silicon transistor may have a maximum allowable operating junction temperature of 200°C. If the case is held at 25°C and the thermal resistance to the case is, say, 0.5 mW/°C, then the maximum power that the unit may safely dissipate is

$$(T_j - T_c)/\Theta_{jc} = P_d, \text{ or}$$
$$(200 - 25)/0.5 = 350 \text{ mW}$$

where T_j = maximum allowable junction temperature—°C

   T_c = case temperature (or T_a, ambient temperature)—°C

   Θ_c = thermal resistance to case (or Θ_a to ambient)—°C/mW

   P_d = allowable power dissipation—mW.

If the transistor's case temperature rises above 25°C, the power that the transistor may safely dissipate must be reduced. For a case temperature of 80°C,

$$(200 - 80)/0.5 = 240 \text{ mW}.$$

Higher junction temperatures lead to higher reverse current, which in turn can lead to more heating. This behavior is called *thermal regeneration*. Obviously such action, if uncontrolled, can cause improper circuit performance such as signal clipping and distortion as the characteristics curves of the transistor rise so that the circuit operates in nonlinear portions of the curves. The most serious result of thermal regeneration, however, is thermal runaway, where the temperature of the junction keeps rising until the transistor is destroyed.

Thermal regeneration is controllable however. The usual amplifier circuit's load resistor is the first line of defense. Since the power dissipated in a transistor is mostly equal to E_c times I_c, we can see from Figure 5-3 that as an amplifier's collector current rises to the operating point of the load line, the junction power increases. But though I_c increases beyond the operating point, the decrease in E_c now reduces the power dissipated to near zero as E_c approaches zero. This is a thermally degenerative region. The load resistor tends to limit the power dissipation in the transistor to a maximum at the operating point.

The secrets of biasing for stability

In most circuits, the control of the temperature sensitivity has more stringent requirements and purposes than just preventing thermal runaway. In Chapter 4, Figure 4-9, a voltage-divider in the biasing circuit for a common-emitter amplifier carries so much more current than the base current that temperature-caused changes in the base current have very little influence on the base voltage. Since the dc base

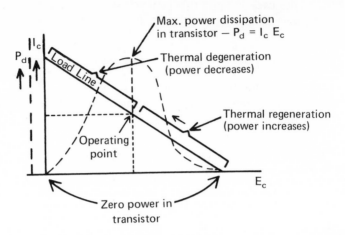

Figure 5-3. A resistor load in the transistor's collector tends to control thermal runaway.

voltage can change very little (as if the base was grounded), the common-emitter amplifier then approaches the lower temperature sensitivity of a common-base amplifier.

In addition, in Figure 4-9, the 1-kΩ emitter resistor provides negative feedback for slow variations, as caused by heating. A slow change in emitter current is counteracted by the resultant emitter-base voltage change across the 1-kΩ resistor. For rapid ac signals, the capacitor that connects across the 1-kΩ resistor provides a low-impedance path to reduce this negative feedback for ac voltages.

Another way to achieve negative feedback for slow, or dc, variations and thus further stabilize the circuit is shown in Figure 5-4(a). In this configuration, any increase in collector current causes the collector voltage to decrease; the voltage output of the bias voltage divider also decreases. A reduced base bias voltage lowers the base current and tends to oppose the original rise in collector current. However, this method by itself is seldom sufficient. For improved stability, the three approaches are combined into one circuit in Figure 5-4(b). Note that capacitor C_b serves a similar purpose as C_e—to bypass ac signals and reduce negative feedback for ac signals.

Emitter-to-base voltage varies with temperature

Another temperature-affected parameter that can be a problem in some applications is the voltage drop, V_{be}, in the forward-biased

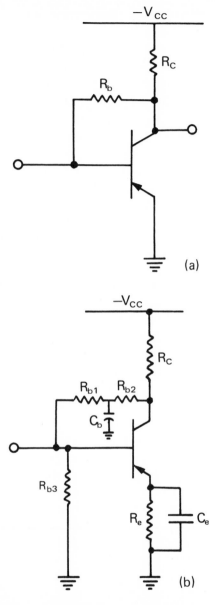

Figure 5-4. Negative feedback from collector to base (a) tends to control thermal problems in transistors, but a combination of collector and emitter feedback resistors (b) does a more effective control job. Capacitors C_b and C_e reduce the usually undesired feedback at the audio frequencies, but do not affect feedback for dc drifts, which temperature changes produce.

emitter-to-base path. In transistors (and diodes), the forward-current threshold, or knee, of the voltage-current curve, where the base current starts to rise rapidly (Figure 3-6), is about 0.6 V for silicon and about 0.2 V for germanium at room temperature. At 100°C the values might be halved to roughly 0.3 and 0.1 V respectively.

If base bias is supplied by a circuit that is essentially a constant current, as from a high-resistance source, this small voltage variation can cause very little change in the base current. Nevertheless, the use of a high series resistor (Figure 5-2(b)) to bias the base is not a good solution because base current varies independently of V_{be}. The current, is strongly temperature dependent. A battery or other low-resistance bias circuit would produce unacceptable bias-current variations because then V_{be} variations become significant. However, the biasing method described in Figure 5-4(b) can resist both these variables.

Simplifying circuit design

Manufacturers' tolerances usually allow a wide spread in initial transistor parameters. Aging also causes shifts—generally decreases in β and increases in I_{cbo}. In modern transistors, though, changes with age have been reduced to a few percentage points. Exposure to nuclear radiation or x-rays greatly multiplies this aging process. Such wide tolerances, then, do not generally justify exact solutions to most transistor circuit designs.

Further, it should now be evident that an exact analysis of even a simple transistor amplifier circuit which tries to take into account every circuit element, tolerance and source of circuit drift is a very complex undertaking, and the resultant answers are even harder to use.

A simple rule-of-thumb approach is often just as good as all the complex formulas. The secret of good transistor circuit design lies in the use of lots of negative feedback to minimize transistor variability. An ideal circuit would work with almost any transistor in a broad range. The only requirement would be that its β is above a certain minimum, and that it can withstand the circuit power requirements.

Figure 5-5 shows such a simplified approach to designing a transistor amplifier stage. Transistor bias is arranged to take advantage of negative-feedback stabilization. Bypass capacitors (shown dotted) to eliminate this feedback for the signal frequencies will be discussed in detail later.

All the simplifying assumptions and principles used in the calcula-

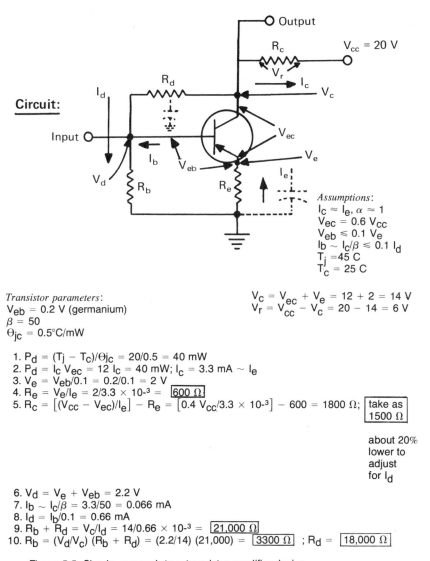

Circuit:

Assumptions:
$I_c \approx I_e, \alpha \approx 1$
$V_{ec} = 0.6 V_{cc}$
$V_{eb} \leqslant 0.1 V_e$
$I_b \sim I_c/\beta \leqslant 0.1 I_d$
$T_j = 45\ C$
$T_c = 25\ C$

Transistor parameters:
$V_{eb} = 0.2$ V (germanium)
$\beta = 50$
$\Theta_{jc} = 0.5°C/mW$

$V_c = V_{ec} + V_e = 12 + 2 = 14$ V
$V_r = V_{cc} - V_c = 20 - 14 = 6$ V

1. $P_d = (T_j - T_c)/\Theta_{jc} = 20/0.5 = 40$ mW
2. $P_d = I_c V_{ec} = 12\ I_c = 40$ mW; $I_c = 3.3$ mA $\sim I_e$
3. $V_e = V_{eb}/0.1 = 0.2/0.1 = 2$ V
4. $R_e = V_e/I_e = 2/3.3 \times 10^{-3} = \boxed{600\ \Omega}$
5. $R_c = [(V_{cc} - V_{ec})/I_e] - R_e = [0.4\ V_{cc}/3.3 \times 10^{-3}] - 600 = 1800\ \Omega$; $\boxed{\text{take as } 1500\ \Omega}$

about 20%
lower to
adjust
for I_d

6. $V_d = V_e + V_{eb} = 2.2$ V
7. $I_b \sim I_c/\beta = 3.3/50 = 0.066$ mA
8. $I_d = I_b/0.1 = 0.66$ mA
9. $R_b + R_d = V_c/I_d = 14/0.66 \times 10^{-3} = \boxed{21,000\ \Omega}$
10. $R_b = (V_d/V_c)(R_b + R_d) = (2.2/14)(21,000) = \boxed{3300\ \Omega}$; $R_d = \boxed{18,000\ \Omega}$

Figure 5-5. Simple approach to a transistor amplifier design.

tions shown in Figure 5-5 have been previously described. The assumptions, or rules of thumb, are listed and typical nominal germanium transistor parameters are used. Power dissipation is assumed to be 40 mW; this limits the case temperature rise to 45°C when the thermal resistance, Θ_{jc}, is 0.5°C/mW and the ambient is 25°C. Emitter-to-base voltage, V_{eb}, is taken as 0.2 V for germanium, and for stability, this voltage should be 10 percent or less of the voltage drop V_e across resistor R_e. Base current I_b is roughly estimated from I_c/β; it should also be 10 percent or less than the bias-voltage-divider current I_d for stability purposes. Both of these percentages have been arbitrarily set at 10 percent. They can be decreased for greater stability, but at a price.

A lower percentage for calculating V_e would require a higher V_e. This would be at the expense of a lower voltage across R_c (V_{cc} is fixed) and would produce a lower amplifier gain, as we shall soon see. A lower percentage for calculating I_d would result in a higher I_d, higher power losses in the bias circuit, the need for a lower R_c and, again, lower gain.

Finally, V_{ec} is proportioned at approximately 0.6 times the given V_{cc}. This proportion can also be varied, but 0.6 places the amplifier's operating point at about the correct point on the load line for the maximum linear output-voltage swing.

Gain calculations made easy

Not many engineers realize that the voltage gain of a common-emitter or common-base amplifier depends primarily on the dc voltage, V_r, across the load resistor R_c. If in Figure 5-5 the loading effects of R_b and R_d are neglected and the feedback effects of R_b and R_e are considered perfectly bypassed for medium-frequency operation, then the amplifier's gain is approximated by

$$G_v = \alpha\ R_c/R_e \text{ (Chapter 4, p. 70)}$$
$$= I_c R_c/I_e R_e$$
$$= V_r/0.0256$$
$$\text{or } G_v = 38.4\ V_r$$

Another way of saying this is that the voltage gain is 38.4 for each volt dropped across R_c; then for $V_r = 6$ V, $G_v = 230$.

In the bias-circuit calculation, it was noted that if the V_e is made higher, it would be at the expense of V_r. The result would be lower gain, as we can now clearly see.

The practicality of this equation is demonstrated in Figure 5-6, where calculated and actual gains are plotted for the circuit of Figure 5-5. The roughly 15 percent deviation at the higher load voltage levels results mainly from the loading of the bias voltage-divider network. Of course, the load of the next stage across R_c, and any imperfect by-

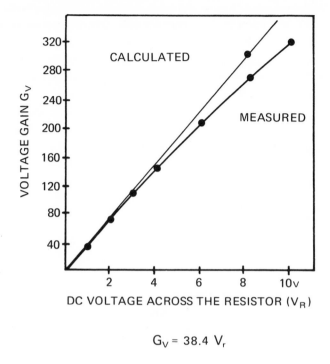

$$G_V = 38.4\ V_r$$

Figure 5-6. Measured values of gain agree closely with values calculated from the simplified equation.

passing of R_e and R_d, will also produce a lower gain.

Simple tests for temperature stability

All transistor circuits, especially power amplifiers, should be tested for temperature stability. Even crude tests are very helpful. A simple way to determine a transistor's (or any other component's) temperature is to touch it—carefully, of course. At 50° C, the little finger (most sensitive) can be held to the part indefinitely. At 60° C, most fingers feel uncomfortably hot after about one second. The case temperature of 60 ° C is generally considered a safe operating temperature for most germanium power amplifier transistors. Silicon transistors can safely withstand over 100° C. At 100° C, water dropped on the transistor housing will sizzle.

To determine how close a circuit is working to its upper temperature limit, a hot soldering iron can provide some external heating. Bring the iron near the transistor and observe the circuit's performance, but be careful not to overheat germanium transistors. If a small increase in transistor temperature causes a marked decrease in the circuit's performance, the circuit is operating too near its upper limit.

At the cold end of the temperature scale, Freon, available in aerosol cans, can be sprayed onto the transistor. This will quickly and temporarily reduce its temperature. Again, the circuit's performance is watched for deterioration or improvement, as the transistor returns to its normal temperature.

Taking frequency into account

In all the circuit configurations that were previously described, the frequency of the applied signal, except for brief mention, was ignored. Although it was not specifically stated, it was generally assumed that the signal frequency was confined to the audio band. The high-frequency performance of a transistor circuit is strongly influenced by the type of transistor used. Even if transistors have the same β and the same current gain at audio frequencies, different transistors will provide widely different responses at high frequencies.

For example, the simplified common-emitter circuit of Figure 5-7(a), when outfitted with one of the older 2N104 transistors with a β of about 50, might provide a relatively flat gain to about 14 kHz, where the gain would fall off to about 70.7 percent of the midfrequency range.

Substitute a high-frequency transistor, say a 2N1177, with exactly the same β (50), and the audio-frequency gain remains the same, but the 70-percent fall-off point will not be reached until possibly 2800 kHz.

The transistor parameter that determines a common-emitter circuit's frequency response is called the *beta cutoff frequency*, f_β.

Much improved high-frequency performance can be obtained with a common-base circuit (Figure 5-7(b)). In this arrangement, the 2N104's 70.7-percent cutoff frequency becomes 700 kHz and the 2N1177 becomes 140 MHz. However, note that the bandwidth improvement of the common-base circuit was obtained at the expense of current gain. The common-base circuit provides somewhat less than a current gain of one (Figure 5-7(c)).

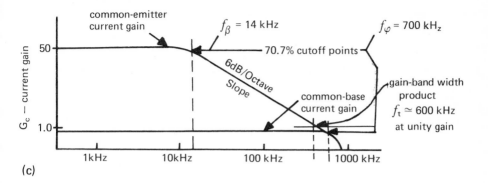

(c)

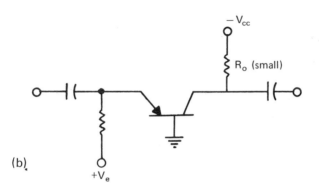

(b)

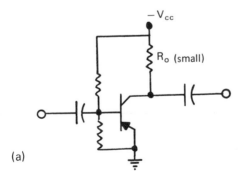

(a)

Figure 5-7. The relative high-frequency cutoff points for common-emitter (a) and common-base (b) configurations are shown in current-gain curve (c) as the alpha and beta cutoff frequencies $f\alpha$ and $f\beta$. By extending a 6 dB/octave-sloped line from $f\beta$, the *gain-bandwidth product*, or frequency at unity gain f_t, is obtained for the common-emitter circuit.

The frequency at which the common-base circuit current gain drops to 70 percent of its medium frequency gain is called *alpha cutoff frequency*, f_α. This frequency parameter is the one most often provided in manufacturers' data sheets and serves primarily as a figure of merit of the high-frequency proficiency of the transistor. It is approximately related to f_β by the equation:

$$f_\alpha = \beta f_\beta$$

Another figure of merit often used is the *gain-bandwidth product*, f_t. It is defined as the frequency at which the common-emitter current gain becomes unity; its value for modern transistors varies between f_α and about $0.5\ f_\alpha$.

Pn junction capacitance limits frequency response

To understand why a transistor exhibits these cutoff frequencies, first refresh your memory about the capacitive action of pn junctions in Chapter 3. Since the base-collector junction is usually reversed biased, a depletion-type capacitance forms between these transistor elements. Then, because the base-emitter junction is usually forward biased, a diffusion capacitor forms.

In Chapter 4, the gain for an emitter-grounded circuit was calculated without reference to frequency. All transistor capacitance effects were considered negligible. This is true as long as the signal frequencies remain below the previously mentioned cutoff frequencies.

Let's first examine the base-emitter diffusion capacitance. In Figure 5-8(a), this capacitor, C_e, is shown shunting the equivalent input resistance for a common-base circuit, $R_e(1 + \beta)$, as derived in Chapter 4. Resistor r_i represents the resistance of the input circuit and some ohmic and other effects negligible for our present purpose.

At low frequencies, the capacitor can be ignored, as in Chapter 4. When the frequency rises so that the capacitive impedance $1/2\pi f_\beta C_e$ equals $R_e(1 + \beta)$, then the gain of the circuit drops to 70.7 percent of the low frequency gain, if the input resistance, r_i, is considered to be high, or the signal is from a current source. Then

$$f_\beta = 1/2\pi C\ (1 + \beta)R_e$$

Thus for the 2N104, $R_e(1 + \beta) = 25(1 + 50) \approx 1300\ \Omega$, if R_e is taken as about twice the theoretical value of $12.8\ \Omega$ (See Chapter 4) and the results are rounded off.

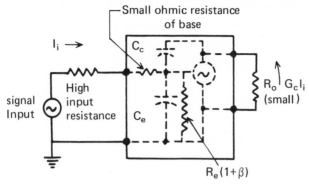

$R_e(1+\beta)$

(a)

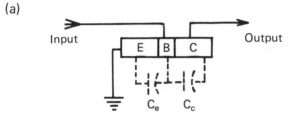

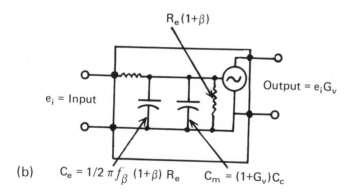

(b) $C_e = 1/2\,\pi f_\beta\,(1+\beta)\,R_e$ $C_m = (1+G_v)C_c$

Figure 5-8. The capacitance between base and emitter, C_e, is represented for a common-emitter circuit as in equivalent diagram (a), when the load resistor R_o is small. In addition, when R_o is large enough to provide appreciable voltage gain G_v, then the base-to-collector capacitance is "reflected" into the input circuit as a much higher capacitor C_m, as represented in (b).

It follows then that

$$C_e = 1/2\pi \times 14 \times 10^3 \times 1.3 \times 10^3 = 8600 \text{ pF}$$

Both the factors $R_e(1 + \beta)$ and C_e form major parts of the approximate equivalent circuit of the actual transistor's input behavior.

In the foregoing discussion of the circuit in Figure 5-7(a) and (b) and their equivalents, the load resistor R_o was clearly labeled "small" to avoid the effects of the base-to-collector capacitance C_c on the circuit. A small load resistor makes the amplifiers operate primarily as current amplifiers with low voltage gain. Appreciable voltage gain introduces the so-called Miller effect, which magnifies the base-collector capacitance as "seen" from the input. The effect is easy to understand. A small voltage change at the base is amplified and of opposite polarity at the collector. This amplified voltage is thus in series aiding with the input signal and results in more input current into the base-collector capacitor than if no voltage gain existed. The base-collector capacitance then appears to be much larger than its C_c value without the Miller effect. Thus:

$$C_m = (1 + G_v)C_c$$

For a voltage gain, G_v, of nine, the base-collector capacitance for a 2N104, which is listed as a nominal 40 pF, would appear as 400 pF in the input circuit (Figure 5-8(b)).

Over-all, C_m (m for Miller effect) combines with C_e to provide a total input capacitance effect of 9000 pF to further lower the high-frequency cutoff point. The 2N1177, which is specially designed for high-frequency operation, has a much smaller base-to-collector capacitance of 2 pF.

It should be remembered that both these capacitor effects vary with the collector-base and emitter-base junction voltages and increase as the junction voltages increase. The capacitance values used in the equivalent circuits should be considered as only nominal. While to assume them constant is useful for an approximate understanding of the high-frequency transistor performance, the nonlinear behavior of transistor junctions is far more complex than these simple explanations might suggest.

Operating at cutoff and saturation

Most of the transistor applications discussed so far have been for linear operation and the so-called "active" region of a transistor's characteristics were used (Figure 5-9(a)). In the active region, the emitter-base junction is forward biased and the collector-base junction is reverse biased.

However, transistor circuits that are used for switching usually operate in cutoff or saturation only, and the active region between them is crossed as rapidly as the transistor and its circuit will permit in switching between the two. In cutoff, both transistor junctions are reverse biased, and in saturation, both are forward biased.

Cutoff (and active) transistor characteristics have been covered previously; the saturation region's properties present some new problems.

A basic saturated transistor switching circuit is shown in Figure 5-9(b); its voltage and current waveforms under typical base-drive conditions are shown in Figure 5-9(c).

At the input signal's lower logic level, the transistor is in a hard cutoff condition because of the reverse-bias voltage, V_{bb}. Only a negligible reverse leakage current flows into the base and the collector current is also very low. The output voltage is therefore V_{cc} "seen" through R_c.

When the input signal rises sharply to its high logic level, base current starts to flow, but the collector current does not immediately follow this rise; it lags behind by an amount of *delay time*, t_d. This is mainly caused by the emitter capacitance, C_e (Figure 5-8), which is also very much present in a transistor used in saturation-mode switching.

As soon as the transistor has made a transition from cutoff to the active region and the collector current starts to increase, the collector-

Figure 5-9. (a) Transistor switching circuits.

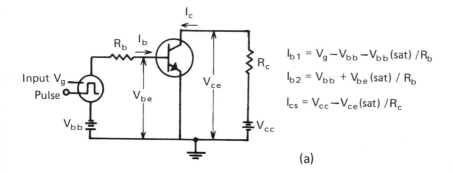

$$I_{b1} = V_g - V_{bb} - V_{bb}(sat) / R_b$$

$$I_{b2} = V_{bb} + V_{be}(sat) / R_b$$

$$I_{cs} = V_{cc} - V_{ce}(sat) / R_c$$

(a)

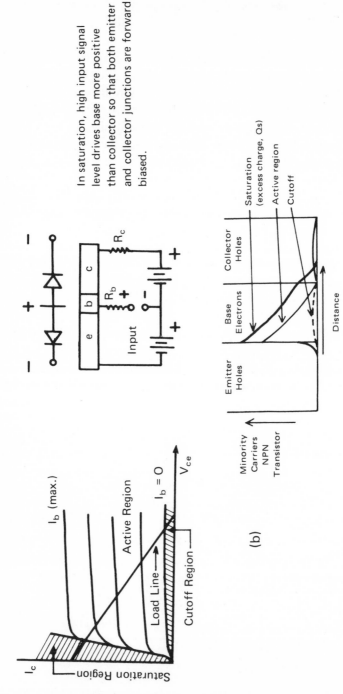

In saturation, high input signal level drives base more positive than collector so that both emitter and collector junctions are forward biased.

(b)

Figure 5-9 (cont.). Transistor switching circuits (a) usually operate in the saturation and cutoff regions of the transistor's characteristics curves (b).

114

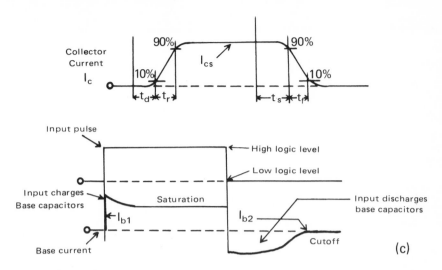

Figure 5-9 (cont.). The only time spent in the active region is during intervals t_r and t_f shown on curves (c).

base capacitance, C_c, becomes effective in controlling the collector current rise, because now the Miller effect can operate. Capacitance C_c causes the collector current to rise ''gradually'' during the leading-edge *rise time*, t_r.

When the input pulse returns to its low level, again the collector current does not change immediately. Excess storage charge, Q_s, that had been forced into the base capacitor must first discharge. The time required for this charge to decay is called the *storage time*, t_s; this time depends mostly upon how strongly the transistor was originally forced into the saturation region by turn-on base current, I_{b_1}. The time interval, t_s, distorts the input pulse, and the output pulse is longer by this amount.

Once the stored charge decays to the edge of the saturation region, the transistor again enters the active region and then the collector current can start to decrease. This fall time, t_f, through the active region is similar to the rise time, t_r, but while the rise time depended upon I_{b_1} (turn-on base current), the fall time depends upon I_{b_2} (turn-off base current) (Figure 5-9(c)).

Because transistor voltage swings in switching operation are large and the capacitor values of C_e and C_c both vary with this voltage, exact capacitor values cannot be assigned to them. An accurate calculation of all the time lags that are based upon C_e and C_c is therefore very difficult to make. Furthermore, to obtain even the nonlinear characteristics of these capacitors is no simple task. The time intervals are therefore usually measured values.

Speed up a saturated circuit

For a saturated logic circuit, the "ideal" driving-current waveform should provide a high initial peak to overdrive the base and charge C_e and C_c rapidly, and then fall to a level to just maintain saturation. The lowered base current prevents excess accumulation of stored charge that slows the turn-off.

Fast turn-off is also assured by a large temporary reverse overdrive current. After the transistor has turned off, the base current should be close to zero—only sufficiently large to keep the transistor cutoff. A low current to maintain cutoff is desirable, because turn-on would be delayed by any accumulation of reverse charge in the input capacitance, C_e.

The normal rectangular logic pulse can be converted to a type more suited for driving saturated transistors by placing a "speed-up" capacitor, C_b, across the base resistor, R_b, as in Figure 5-10(a). Figure 5-10(b) shows the current waveform into the base that is created by this speed-up capacitor. Note the similarity of this waveform to the described "ideal" waveform.

The proper value for the speed-up capacitor, C_b, is best (and most easily) determined experimentally. This is done by observing the trailing, or turn-off, edge of logic signal outputs from a logic gate as C_b is increased. The correct value of C_b has been reached when its further increase shows little or no further sharpening of the output waveshape's trailing edge (Figure 5-10(c)).

For high speed, don't saturate

Since the problem of storage time does not arise if the transistor is not allowed to saturate, a substantial improvement in switching speed can be obtained by the use of nonsaturating switching circuits.

One way to prevent saturation is to clamp the collector so that the collector-base junction is prevented from becoming forward biased (Figure 5-11). If the voltage drop across R_{b_2} is kept larger than the forward drop across the clamping diode, then the base collector junction can never become forward biased. Base current, I_b, above the desired level into the base, will flow through the clamp diode.

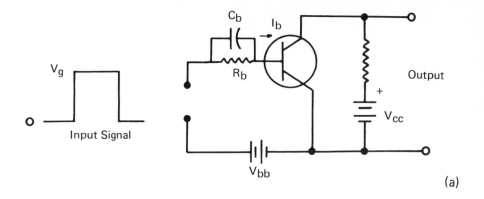

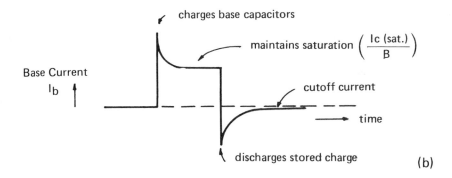

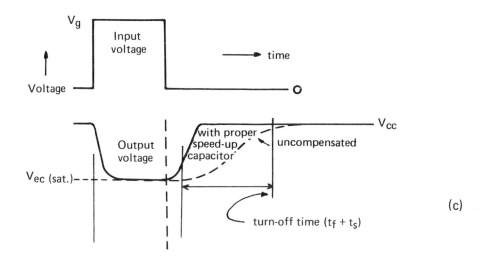

Figure 5-10. A logic circuit (a) that works into the saturation region uses a speed-up capacitor C_b to modify the base-current waveform as in (b) so that the turn-off time is minimized (c).

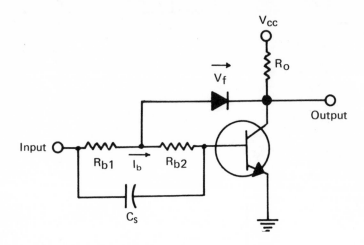

Figure 5-11. A clamping diode will prevent the transistor from becoming forward biased if $I_b R_{b_2} \geq V_f$, where V_f is the forward drop across the diode.

A speed-up capacitor can also be used in this circuit, but it can be much smaller than for a saturated circuit. It doesn't have to handle any excess charge; it merely has to bypass R_{b_1} and R_{b_2}.

Many other nonsaturation switching circuits are used and they generally can provide from two to five times the speed of the highest speed saturated types, but invariably the designer must pay a price of increased transistor power dissipation and more costly and complicated circuits.

Chapter 6

Working with Devices That See and Shine

In the opening chapters, semiconductors are characterized by a forbidden energy gap that valence electrons must bridge to become conduction electrons. The energy of the heat motion of the atoms can propel a very few of the valence electrons into the conduction bands. Higher temperatures mean higher energies; thus the number of electrons that jump the gap increases with temperature, and a semiconductor material's conductivity increases.

When a valence electron jumps to the conduction band, a hole is left behind and an electron-hole pair of minority carriers, as previously defined, is created.

Not only heat, however, but also radiant or "light" energy that strikes an atom can cause a valence electron to jump to the conduction band. The word "light" as used here is not limited to the part of the electromagnetic spectrum that the eye can see, but it can include any wavelength that "illuminates" the semiconductor material as long as it has an energy greater than the jump required by the forbidden gap, Eg. Only a relatively narrow band of wavelengths can be seen with the human eye—about 0.38 to 0.78 microns (μm). The complete spectrum extends from long radio waves to gamma rays (Figure 6-1).

Since light can be considered as particles with energy content, E = hf, where h is Planck's constant and f is the light's frequency, then hf must be greater than E_g to generate electron-hole pairs.[1]

---

[1] See pages 28 to 29.

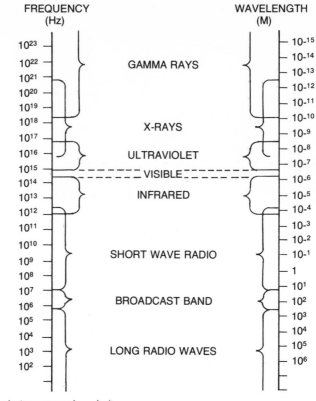

E = hf—photon energy in = hc/λ
λ = c/f = wavelenght = velocity/frequency
$E = 12.4 \times 10^3/\lambda(\text{Å}) = \dfrac{6.63 \times 10^{-34} \times 3 \times 10^8}{10^{-10} \times 1.6 \times 10^{-19}}$ electron-volts
I Angstrom (Å) = 10^{-10} m = 10^{-4} μm (microns)
c = velocity of light = 3×10^8 m/s
h = Planck's constant = 6.63×10^{-34} joule-sec
I joule = 1.6×10^{-19} electron-volts

Figure 6-1. Visible light is only a small portion of the total electromagnetic spectrum.

When semiconductor material is suitably arranged for excitation by light energy, it is called a sensor or detector of radiation. If the material acts as a sensor, light illuminating its surface will be absorbed and electron-hole pairs will be generated to add to the thermally derived minority carriers. If, however, the photons of light have energy less than the forbidden gap, then the material is transparent to the light. A good example is the material germanium, opaque to visible light, but transparent (and insensitive) to infrared.

On the other hand, when minority carriers recombine—a conduction electron falls directly into a hole and becomes a valence electron—energy is released. Depending upon the material and the levels of energy involved, heat or light is produced. If light is generated and can get outside the material, the arrangement is called an emitter.

Both photosensitivity and photoemission can occur in bulk semiconductor material. As will be explained later, however, a diode configuration can produce improved results. Photosensitive devices also often have bipolar and FET-transistor configurations.

Optoelectronics deals with photoemitters and photosensitive devices. A host of components for applications from card readers to laser surveying instruments have been developed. Optoelectronic devices come in a wide variety of types, sizes, shapes and packages. We will examine photosensitive devices first.

Photoconduction in semiconductors

There are two primary effects that govern the operation of solid-state photosensitive devices: photoconduction and photogeneration. Photoconduction is the more widely used effect for control applications. Photoconductive cells existed long before quantum mechanics was applied to solid-state devices. Selenium was one of the first materials discovered to have a photoconductive effect. The first photoconductive cells were constructed of "bulk" materials. Modern bulk material photocells like cadmium sulfide (CdS), cadmium selenide (CdSe) and lead sulfide (PbS) work because their conductivity varies with illumination. They are made from powders sintered on a ceramic substrate (Figure 6-2) to form an essentially uniform conducting mass, with megohms of resistance in the dark, which in light drops to the thousands or even tens of ohms in large units.

Figure 6-3(a) shows the minimum frequency (longest wavelength) of the spectral response of a number of common semiconductor materials in wavelength λ measured in microns (μm $= 10^{-6}$ meters) and the relationship to the material's energy gap, E_g, in electron volts. If hf (Planck's constant times frequency) is less than E_g, the photons will not be absorbed. On the other hand, if hf is high, then the photon is likely to be absorbed near the surface of the photodevice, and the hole-electron pairs that the photon generates never enter into the current flow.

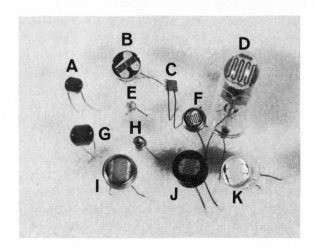

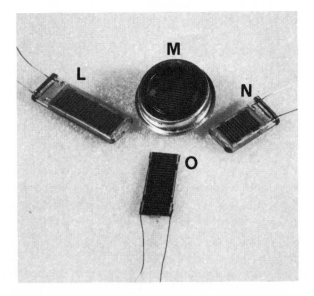

Credits— A,G—Vactec—exposure control
        B,E—Clairex—exposure control and projector auto focus
        C,F,H,J—Clairex—general purpose
        I—Clairex—control street lights
        D,L,M,N,O—Amperex—power photocells handle to 500 mA at 400 V.

Figure 6-2. Bulk-semiconductor, variable-conductivity photocells such as these cadmium sulfide and selenide units (a), and larger power units (b), find a large variety of applications—from turning street lights on and off to controlling film exposure in cameras.

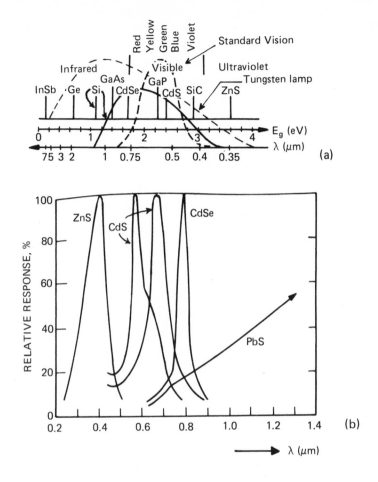

Figure 6-3. A photocell's spectral response is primarily determined by its material's energy gap.

Silicon is a particularly favored material because its broad spectrum begins in the near infrared region and covers the whole visible spectrum. Its wide use results from low cost and a well-developed material technology. Its infrared response drops to near zero at wavelengths longer than about 1.2 μm. Depending upon the structure of the particular unit, a typical short wavelength cutoff is about 0.4 μm, at the upper or violet end of the visible spectrum. Figure 6-3(a) also shows the spectral spread of normal vision and a tungsten incandescent lamp for comparison. The broad overlap of the visible tungsten lamp and silicon spectrum accounts, in large measure, for the wide use of silicon in so many applications. Cadmium sulfide can be peaked in the visible region, and cadmium selenide in the red end of the visible.

Other semiconductor materials can be used with energy gaps to match almost any desired frequency spectrum. For instance, zinc sulfide (ZnS) operates in the ultraviolet region and lead sulfide in the infrared (Figure 6-3(b)). Impurities in the semiconductor often provide sensitivity to frequencies lower than the band-gap requirements of the main part of the bulk material.

Bulk photoconductive cells are slow

The response time of photoconductive cells to light pulses is limited by the time it takes for the current carriers to drift through the device under the applied electric field and the lifetime of the light generated hole-electron pairs. Too short a lifetime prevents the current carriers from contributing to the conductivity, but too long a lifetime limits the rate at which the cell can follow light pulses.

The lifetime of these optically generated electron-hole pairs, before they meet oppositely charged particles and are neutralized, must be sufficiently long to allow them to become captured by the applied electric field. Otherwise they can't contribute to electrical conduction. The device's geometry can be designed to increase response speed, but this is often done at the expense of sensitivity or other desirable properties.

A short conduction path will reduce drift time and increase response speed; however, the reduced illumination area decreases sensitivity. Also, a short path does not allow a high dark resistance—a usually desirable characteristic.

To increase the dark resistance, photocell material is deposited in a folded-line or zig-zag pattern which increases the active length of the cell (Figure 6-2). This also increases sensitivity. Dark-to-light ratios of several orders of magnitude are quite common (100 to 10,000:1).

The history of the way illumination is applied to a bulk conductivity cell affects its subsequent response to light. Exposure to strong light over a long period of time decreases a cell's sensitivity and increases its resistance. Even for short exposure, these cells tend to exhibit such a hysteresis effect but to a lesser degree. Cadmium sulfide cells are particularly prone to this effect, the selenides somewhat less. Such cells, therefore, may not be the best choice for the continuous monitoring of light intensities; however, intermittent measurements, with a chance for the cell to "rest" in between measurements, would be a better application.

Since these bulk cells are basically semiconductor materials, temperature affects them like any semiconductor. Dark currents increase with temperature. Temperature effects are most pronounced at low illumination levels. Sensitivity is decreased with an increase in temperature.

Photoconductive bulk cells in intermediate sizes can operate sensitive relays without amplifiers; large power cells can dissipate as much as 3 W with the help of a sink and handle 0.5 A (Figure 6-2(b)).

Bulk photoconductive cells usually have fairly low frequency responses, besides suffering from hysteresis or memory effects. The responses range from a few hundred to the low thousands of cycles per second. Nevertheless, photoconductors are particularly useful where the offset voltages of other cells cannot be tolerated, such as in photo-chopper applications. Photocells that are designed around a pn junction can have speeds into the megacycle region.

Junction photocells are fast

Solid-state pn junctions, in addition to responding to directly applied voltages and currents in useful ways, also can react to external stimuli such as light (see Chapter 2). Conversely, some specially constructed diodes can emit light.

Let's first examine how light interacts on a pn junction. A plot of reverse bias current of a photosensitive diode (Figure 6-4(a)) shows that the reverse saturation current of the junction increases from its minimum thermally generated value in proportion to the level of illumination. Note that the reverse current is essentially independent of the applied voltage.

Depending upon its application, a photodiode may be operated in a photoconductive mode, as in the third quadrant of the I-V curve of Figure 6-3(b) or in a photogenerative, or photovoltaic, mode, as in the fourth quadrant. In the photovoltaic mode, power can be drawn from the cell. In the photoconductive mode, the cell absorbs power. The so-called solar cells are power versions of photovoltaic cells, but both generative and conductive are used as photodetectors for measuring illumination levels and for converting time-variant optical signals to electrical signals.

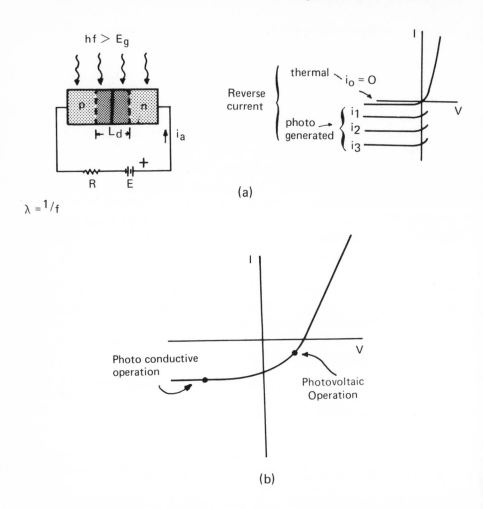

Figure 6-4. A photodiode can be operated with external power (a), in a conductive mode, or in the fourth quadrant (b), a photovoltaic mode, to supply power to an external load, as in solar cell applications.

The response of a photosensitive cell to a square pulse of light is similar to the response of an RC network to a square wave (Figure 6-7(c)). Both have delays in their rise and fall times, t_r and t_f. Low levels of illumination and high applied cell voltage produce long rise and fall times; thus for highest response rate, illumination should be high and voltage low.

The conductive mode is best for high-speed response. For a cell to respond in, say 1 μs, the hole-electron pairs generated by the illumination must reach and sweep across the depletion area, L_d, of the junction in much less than 1 μs (Figure 6-4(a)). To reduce the time, the illumination should be concentrated as close to L_d as possible. Better yet, L_d should be made large enough so that most of the photons are formed within this region.

When the carriers are generated mostly within the depletion area, the cell is called a *depletion-layer diode*. Specially designed diode junctions with a layer of intrinsic (undoped) semiconductor material between their n and p layers, provide a way for producing a wide L_d region. The "i" layer of such a p-i-n junction need not be truly intrinsic but it should have a high resistivity. When the junction is reverse biased, most of the applied voltage appears across the "i" layer. If the hole-emitter pair lifetimes in this i region are long compared to their drift time, most of them will be collected by the p and n layers. Since this can be done in a very short time, the photodiode's response can be quite rapid. On the other hand, L_d must not be so wide that excessive time is required for the light-generated carriers to drift out of the depletion area. Such a long drift time, often called storage time, reduces the cell's response speed by prolonging the fall time of an output pulse.

Another advantage of a moderately wide L_d is that the capacitance of the junction is less. The resultant photodiode's smaller RC time constant with its external circuit then permits the circuit to respond more rapidly.

Silicon planar p-i-n photodiodes are available with very low dark currents; they provide very fast response to pulsed radiation—from less than one to a few nanoseconds. Because the dark currents are less than a few hundred picoamperes, these cells can be used for the detection and demodulation of very low light levels, such as occur in star trackers. In some applications they rival photomultipliers.

Some p-i-n diodes are designed to operate in an avalanche mode. Such *avalanche photodiodes* provide extremely high sensitivity because of the avalanche multiplication.

More sensitivity with phototransistors

If a reverse-biased, base-collector junction of a transistor is illuminated, electron-hole pairs are created as in a photodiode; however, the β gain of the transistor multiplies the base current, as in an ordinary transistor. The phototransistor is thus many times more sensitive than the simple photodiode. The dark, or collector-to-base, leakage current is also amplified, however, and this is a potential source of problems. On the other hand, silicon npn phototransistors have such low dark currents that often only the emitter and collector leads are brought out (Figure 6-5). Adjustment for dark current is not needed in many applications.

Figure 6-5. Low, dark current allows some phototransistors to use two leads only —the emitter and collector. But three-lead, or base-lead, versions allow more flexibility.

Other types with a third or base lead offer the designer greater circuit flexibility. They allow gain and dark-current adjustment, but generally at a loss in sensitivity over open-base types. Any base-to-emitter resistance for the adjustment circuit shunts some of the current that would otherwise flow through the base and be amplified. In addition, the base lead is very useful for stabilizing the phototransistor over a wide temperature range in the same manner as in an ordinary transistor, and when light signals must be combined with control signals.

The operating characteristics of phototransistors look very similar to that of the plot of the characteristics curve of a common-emitter transistor, except that the input is the level of illumination (Figure 6-6). Collector dark currents are on the order of a few nanoamperes, while light currents typically range in the milliamperes. Most phototransistor speeds are rated in terms of turn-on and turn-off time in the microsecond range.

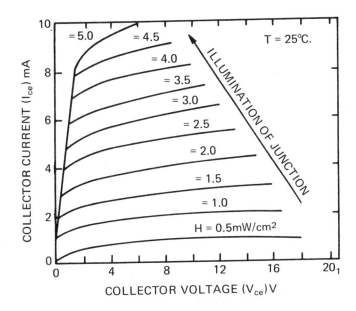

Figure 6-6. A phototransistor's characteristics curves look like ordinary transistor common-emitter curves, except the level of illumination replaces the base current.

For even greater sensitivity and low offset voltage as in bulk photoconductive cells, FET phototransistors, with their inherent high input impedances for the developed photocurrents, can provide five to ten times the current output that the bipolar phototransistor can deliver. A sensitivity of 10 mA/mW/cm² is typical for the FET type vs 1 to 2 mA/mW/cm² for the bipolar.

However, the high input resistance and typical internal gate-to-sink capacitance of about 25 pF results in a time constant of about 25 μs, which is considerably larger than in bipolar phototransistors and can produce slow response. The use of a proper circuit, however, can reduce the effective time constant to about 5 μs. (Figure 6-7(b)).

An operational feedback circuit such as that in Figures 6-7(a) and (b) provides a virtual ground so that the photo device hooks into an essentially short circuit and operates in a shorted mode. This is the best mode of operation for current generated devices such as photodiodes and FETs. This has the effect of making internal capacitances negligible and permits more rapid response. A p-i-n photodiode can achieve rise and fall times of about 0.1 μs and operation between 10 to 20 MHz (Figure 6-6(c)). Photoactivated FETs can have typical rise and fall times of about 0.9 μs in such a circuit.

SCR cells handle high currents

For high output current-carrying capacity, the light-activated silicon controlled rectifier (SCR) can carry forward currents of an ampere or so and handle as high as 50 W with efficiencies near 100 percent. It is analogous to the regular SCR except that light provides the gate triggering. The gate-cathode junction acts as a photodiode to generate the triggering current. Like the regular SCR, once the device is triggered it conducts until the voltage across it is removed or reversed. The units are fast: turn-on times range from somewhat under 1 to about 30 μs, and turn-off is limited by the unit's inherent recovery time of about 30 μs.

The photoactivated SCR, however, is more sensitive to heat than ordinary SCRs. The intensity of illumination needed to trigger the SCR decreases with increased temperature. The required level of light to fire the SCR is thus greatest when the device is first turned on. As the current through the unit warms it up, the amount of illumination needed drops. Increased device voltage and higher bias current also reduce the triggering level; thus for stable triggering, the voltages applied to photo SCRs should be regulated, and load currents should be limited to a few milliamperes.

Light-activated SCRs are used to drive relays; other more powerful SCRs have many industrial control applications.

Sensors that generate power

All the photosensitive devices described so far work on the principle of photoconduction. As noted in Figure 6-4(b), when a photo-

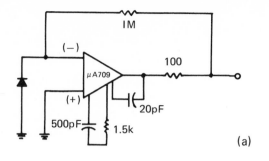

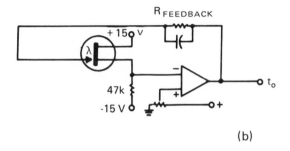

(a)

(b)

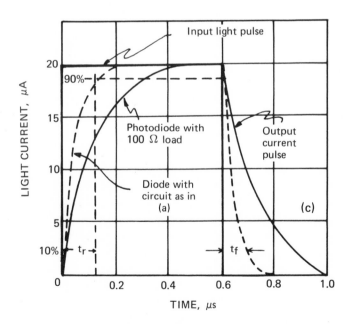

(c)

Figure 6-7. With use in a suitable circuit, a photodiode (a) or photoactivated FET (b) can provide rapid response to input light pulses.

diode is operated in the fourth quadrant of its I-V curve, it can generate power and deliver it to an external load. This is the photogeneration mode; no bias voltages are applied to the unit. Figure 6-8 shows the fourth quadrant reoriented for more convenient use in the manner that a cell manufacturer might present his device's properties. Note that some cells can be operated in either the conductive (third quadrant) or generative mode, and at low illumination, the polarity of the output can be made to reverse.

To obtain a maximum amount of optical energy, photogenerative cells are generally large area devices (Figure 6-9(a)) with their junctions located close to the exposed surface; however, some applications like punched-card and punched-tape readers require small units arranged in closely spaced arrays (Figure 6-9(b)).

Selenium in the photogenerative mode is probably the granddaddy of solid-state cells for light meters and the automatic control of camera exposures; however, selenium generally has been replaced by more sensitive and smaller cadmium sulfide conductive types in camera mechanisms.

From Figure 6-8 it can be noted that the current output from a photogenerative cell tends to be flat with output voltage for a considerable range. This is characteristic of a constant-current generator. When the cell is open circuited, the output voltage follows a logarithmic curve as the illumination varies. At low illumination levels, though, the relationship of voltage to illumination is roughly linear. Except for very small cells, output voltage does not change with cell area for a given type of construction and material.

When current and power are to be drawn from a photogenerative cell, it is important that the dc resistance of the cell be as small as possible so that a minimum of power is lost in this internal resistance. This, of course, is helped by large illumination and junction surfaces and multifingered contacts to increase the area of voltage pickoff from the cell and reduce the resistance, as compared to making contact only at the edges.

The voltages generated by a single cell are generally less than a volt, and the current, a few milliamperes; however, if many cells are connected in series-parallel arrangements, the resultant power can be significant. Such large arrays are used to supply electrical power for space satellites (Figure 6-9(c)).

The frequency response of photogenerative cells tends to be low because of large cell size; however, as the diameter of a cell is reduced,

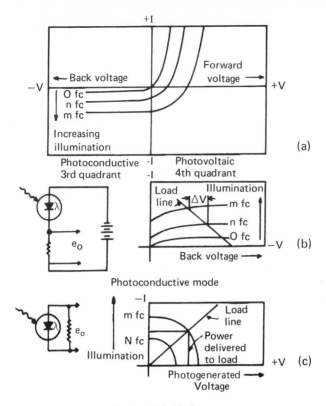

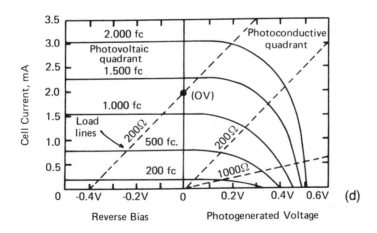

Figure 6-8. A photodiode's performance I-V characteristics (a) are turned around for convenience. The photoconductive mode (b) and photogenerative mode (c) are shown separately, as taken from the third and fourth quadrants of the diode's I-V curve. Some units are designed to work in both modes (d).

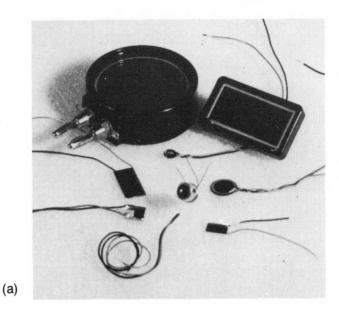

(a)

Figure 6-9. Photogenerative cells come in large surface-area styles (a) for control and illumination measurement, in arrays (b) for reading punched cards and tapes, and in large panels (c) for power generation on space satellites.

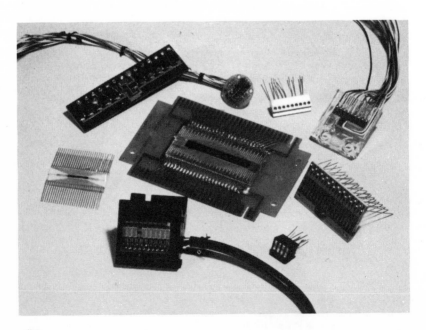

(b)

(c)   Photo courtesy of RCA.

a typical silicon cell can exceed a response rate of 15 kHz. With proper load matching and neutralization of the cell's capacitive effect with a peaking inductor, small cells may approach a 1-MHz rate.

As with any junction device, photogenerative cells are also temperature sensitive. As temperature increases, short-circuit current increases but open-circuit voltage falls. Depending upon the level of illumination and load resistance, the changes can be quite troublesome in many applications.

The diode as a source of light

When the material and construction of a forward-biased diode allow the recombination of electrons and holes directly across the forbidden energy gap, photons (light) are given off with energies that are restricted to a narrow band of frequencies near E_g (Figure 6-10(a)). Such a diode is called a light-emitting diode (LED). Recombinations that are not direct but in many small steps (Figure 6-10(b)) give up their energy mainly in heating the crystal lattice.

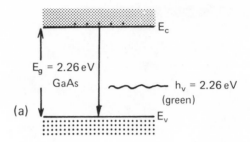

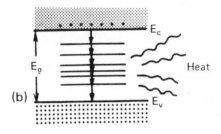

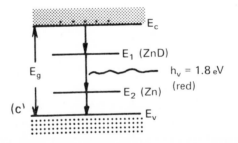

Figure 6-10. Material can be selected where the direct transition across its energy gap, Eg, can produce the desired wavelength of light (a). Too many small intermediate transitions (b) caused by impurities merely heats the crystal, but selected impurities can modify (increase) the wavelength to other desired colors (c).

A wide choice of energy gaps among the various compound semiconductors permits the construction of LEDs that emit for the infrared through the visible and ultraviolet bands (Figure 6-3(a)). Mixed compounds allow further variations. The proportions of As to P in GaAsP (gallium arsenide phosphide) varies the gap from 1.43 eV for GaAs to 2.26 for GaP. This changes the emitted light from infrared to green (Figure 6-11).

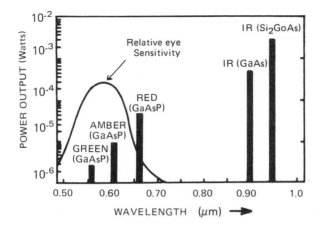

Figure 6-11. Gallium arsenide to gallium phosphide, and a combination of arsenide and phosphide in between, can change the color of emitted light.

	RED	AMBER	YELLOW	GREEN
VOLTS	1.6-2	2-2.8	3-4.5	3-4.5
LUMINOUS EFFECIENCY (%)	3.2-10	50.3	75.7	99.5
POWER EFFICIENCY (%)	0.06-0.1	0.01	(NO DATA)	0.0015
50 FT-L CURRENT (mA)	1	(NO DATA)	5	5

Impurities can also effect considerable variations in color. For instance, *donor* impurities such as Te, S and Se can produce an intermediate step near the conduction energy level roughly 0.1 eV below the condition band, and a ZnO pair, about 0.4 eV. *Acceptor* impurities such as Cd and Zn introduce levels close to the valence levels. If ZnO and Zn dopant impurities are properly introduced into a GaP, a red transition of 1.8 eV is produced (Figure 6-10(c)).

A turn-on, minimum forward-bias voltage of about 1.5 to 2.5 V is required to produce a light output on most LEDs, and the current rises very steeply with applied voltage thereafter (Figure 6-12). To keep the

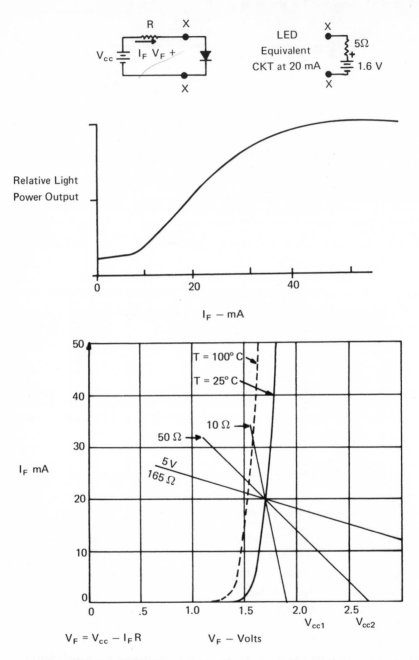

Figure 6-12. A simple series resistor is often sufficient to control the LED's current. It is more effective to control an LED's output by varying its current rather than its voltage. A better match of the outputs of samples of a particular type is obtained at high current levels, say 30 mA, instead of 5 mA.

138

current within safe limits, a constant current source or at least a series resistor is required.

Light output is proportional to current at an LED's mid-current range. At low currents a high portion of the current does not contribute to light output, and at high currents, ohmic drops and heating losses within the diode also represent losses from light production. The light at both low and high currents is therefore not linear.

When an LED heats up, its light output tends to shift to longer wavelengths and its output efficiency falls off significantly. At higher temperatures, minority carriers can recombine more readily. The probability that an electron will not directly drop to a valence level is increased and more of the recombinations thus become nonradiative in nature.

One unique feature of some LEDs is their rapid turn-on and turn-off times. Response times of 5 ns are easily achieved. No other source of light can so easily provide such high-speed light pulses. And because the pulses can be so short, the current and light-intensity peaks can be very large, since the rms current that does the heating is thus low. Peak currents can easily be 10 to 100 times the allowable average current.

LEDs take many shapes

LEDs are available in several categories:

- Panel lamps and indicators—usually single diodes
- Alphanumeric displays—arrays of diodes
- Infrared-emitting diodes for controls—single diodes, nonlasing
- Ranging and other pulsed systems—arrays and single diodes, lasing.

Long life and high light-output efficiency enable LEDs to pose a threat to the future of the incandescent panel lamp. A typical power requirement is about 2.1 V at 10 to 20 mA; lives are given in tens of thousands of hours. Many LEDs are packaged as exact replacements for popular sizes of incandescent lamps (Figure 6-13(a)).

Tiny LEDs arranged to form letters and characters have contributed, in large measure, to the practicality of pocket calculators; even in the larger sizes they compete with all other forms of alphanumeric display methods (Figure 6-13(b)).

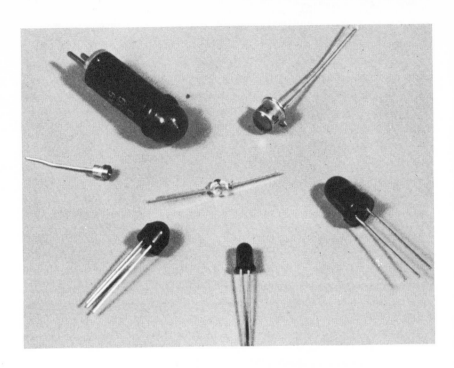

Figure 6-13. Some LEDs, as in (a), can directly replace conventional pilot lamps; others, as in (b), are arranged in arrays to form alphanumerical characters for displays.

Infrared diode LEDs match the sensitivity of silicon diode and transistor detectors. Emitter and detector combinations in arrays read holes in punched cards and tapes (Figure 6-9(b)), and larger individual sets act as intrusion detectors and object counters on production lines.

Emitter diodes with power outputs of 6 to 65-W peak power, and stacks of them that range from 50 to 300-W peak serve in gallium-arsenide LASER systems. For still higher powers, cryogenically cooled units in LASER arrays are sometimes used.

Geometry makes a difference

The geometry of the LED's construction and the lens, if any, has a large effect upon the amount of radiation that is finally emitted after the unit's internal quantum efficiency takes its toll. *Quantum efficiency* is the radiative energy generated divided by the electrical power consumed before any of the LED's sources of power loss are considered. A major problem in LED construction and source of loss is the method of affixing the front electrical contact so that the contact resistance is low, without, at the same time, blocking any significant amount of radiation.

Many arrangements have been devised for extracting the maximum amount of radiation. The most common structure is flat, with the only usable light emitted from the top of the chip (Figure 6-14). Several types use a miniature parabola that collects the edge emission and directs it forward along with the top surface emission. Some edge-emitter types use large top contacts to improve electrical efficiency and depend mainly on the edge emission for the light output.

Though a reflector can significantly improve the performance of an LED, better results can be obtained when the structure allows more of the radiation generated internally within the chip more access to the outside. A substance like GaAs has a high index of refraction (about 3.6) so that radiation that arrives at a flat surface with an angle greater than 16° from the normal is reflected back into the chip. If the chip's active material is shaped like a dome, the light arrives at this surface almost normally everywhere, and very little is reflected back. Even though the material is far from transparent, the efficiency is improved by a factor of ten over a typical flat structure. If the material were perfectly transparent, the improvement would be closer to 25 times.

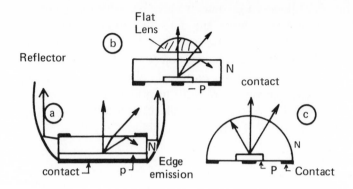

Emitter Structure	Figure Applicable	Power Output (mW)	Collection Optics	Cost
Flat	(a) , (b)	0.5–1.0	Lens	Low
Edge	(a)	0.5–3.0	Mini parabola	Low
Flat/Edge	(a)	3.0	Lens/Mini parabola	Low
Dome	(c)	10.0	parabola	High

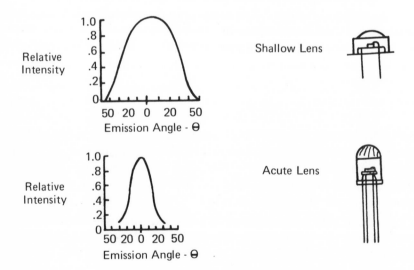

Figure 6-14. The shape of the emission material's structure and the presence of reflectors and lenses affects both the output efficiency and viewing angle of the LED.

Plastic or glass lenses allow the light that does emerge to be concentrated or distributed. A shallow lens allows a broad emission angle and an acute lens, much thicker in the middle than at its edge, provides a narrower beam. A diffuser-type lens is sometimes advantageous. A good diffuser lens will lose not more than 10 percent of the light, but it can greatly enhance the contrast of the emitted light against ambient light. A clear lens will allow the reflection of ambient light; this can often wash out the LED's light output. A colored lens can also help against ambient washout. Also, the light from a LED originates from a relatively small area and is very intense; this could make for difficult or uncomfortable viewing. A diffuser lens spreads the light over a wider area and increases the viewing angle.

Measuring luminance

Electronic specialists are generally not familiar with the techniques and units of light measurement. To work effectively in the optoelectric field requires familiarity with terms like *lumens, candelas* and *lamberts*, and a "feel" for the values that common sources and surfaces radiate and reflect (Figure 6-15).

Because there are at least a half dozen different sets of photometric units, it is very easy to become hopelessly confused, but let's see if we can steer you through the maze.

It's probably easiest to start with the *lumen*, which has the dimensions of power. One watt is equivalent to 68 lumens (lm) at the peak of the human-eye-sensitivity (photometric) color curve (Figure 6-3(a)), or 0.555 μm (green).

The lumen is often referred to as *light flux* that can be spread out to illuminate a surface. It can come from a concentrated source or extended sources of various shapes. When the light energy or flux (F) in lumens comes from a point source, the *intensity* (I) of the source is measured in lumens/steradian (1m/sr). A unit of intensity is called the *candela* (cd) and equals 1 lm/sr.

A steradian is a unit of solid angle. There are 4π steradians to a sphere (Figure 6-16(a)). This is analogous to the radian in two-dimensional geometry, where there are 2π radians per circle (360°).

Because there are $4\pi \sim 12.6$ steradians in a sphere, the total flux that one candela generates is 12.6 lumens, or 4π I lumens for a point source of intensity, I. Of course, we have assumed that a point source

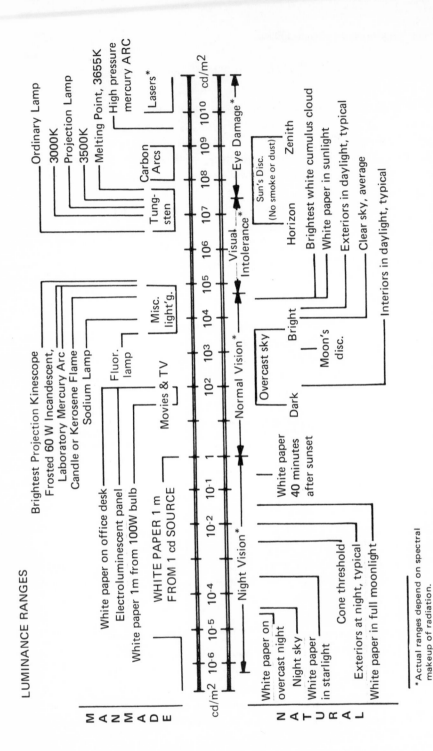

Figure 6-15. This chart can provide the electronic specialist with an idea of the brightness, or luminance, that common sources and surfaces produce.

*Actual ranges depend on spectral makeup of radiation.

144

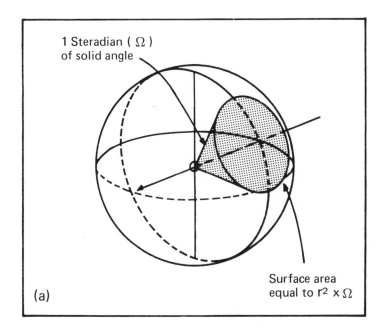

(a)

1 Steradian (Ω) of solid angle

Surface area equal to $r^2 \times \Omega$

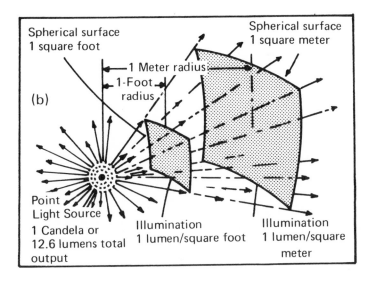

(b)

Spherical surface 1 square foot

Spherical surface 1 square meter

1 Meter radius

1-Foot radius

Point Light Source 1 Candela or 12.6 lumens total output

Illumination 1 lumen/square foot

Illumination 1 lumen/square meter

Figure 6-16. A theoretical point source radiates uniformly over 4π steradians of space.

radiates uniformly in all directions. For a small area, A, that is at right angle to a radius from a point source, the amount of flux that strikes it is approximately

$$F(lumens) \sim A(I/r^2)$$

where r is the distance from source to surface (Figure 6-16 (b)). Flux radiating in all directions, thus, spreads out as the square of the distance from a point source.

While many light sources like LEDs approximate point sources very closely, many bar- and panel-type lights are better described as area sources. The intensity of point-source LEDs are generally specified in candelas, or millicandelas (mcd), but displays and indicators that are area-like need another kind of measure that accounts for their different geometry. Photometric specialists use the *lambert* to describe the *brightness* of area-type sources. The lambert is related to the candela by a constant and area factor to take care of the geometry difference, at least in theory, to make the lambert and candela more comparable on a radiated-energy basis. Note that the word *intensity* is used to describe point sources and *brightness* is reserved for area sources.

Theoretically, a small area is assumed to radiate (or reflect) on one side only and therefore only half as much flux comes out of it when compared to a point source. Another assumption made is that the surface is perfectly diffuse and radiates or reflects according to Lambert's law of cosines–the emission brightness varies as the cosine of the angle from a normal to the area. In practice this is only approximated, but this is a necessary factor in relating point and area source, intensity and brightness units.

Calculus shows that because of Lambert's law, an area source emits half again as much flux as a point source. Only ¼ (4π) steradians of total flux can theoretically come out of a small area source as compared to a point source. Therefore,

$$1 \text{ lambert} = 1 \text{ lumen/sr} \cdot \text{cm}^2) = (1/\pi) (\text{cd/cm}^2) =$$
$$(10{,}000/\pi) (\text{cd/m}^2) = 3183 \text{ cd/m}^2.$$

The cm², or m², dimension is needed to spread the given flux over the source's area (Figure 6-17).

From here on photometric units become a semantic nightmare. We only hinted at this problem when we said that point sources exhibit

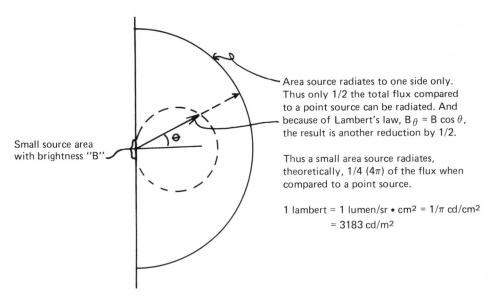

Small source area
with brightness "B"

Area source radiates to one side only.
Thus only 1/2 the total flux compared
to a point source can be radiated. And
because of Lambert's law, $B_\theta = B \cos \theta$,
the result is another reduction by 1/2.

Thus a small area source radiates,
theoretically, 1/4 (4π) of the flux when
compared to a point source.

1 lambert = 1 lumen/sr • cm^2 = 1/π cd/cm^2
= 3183 cd/m^2

Figure 6-17. Area sources and point sources use different luminance units.

intensity but area sources show brightness. The distinction between intensity and brightness is subtle enough, but photometric specialists have "refined" these terms and replaced them with "luminous intensity" and luminance, respectively. Many other confusing terms have been introduced, and now supposedly obsolete terminology exists in the literature side by side with newer and not much clearer words like illuminance and irradiance for illumination (surface brightness from an external source), and obsolete candlepower, which is replaced by the terms luminous intensity and luminance.

The number of equivalent units that exist to describe each quality is beyond belief. For a confusing and only partial table of conversions, refer to Figure 6-18. Units like hefners, nox, carcel units, the English sperm candle, candlepower, etc., abound but have been left off the tables. Only the more modern units are listed.

Fortunately, by relating all units back to lumens or watts, you can usually work your way out of any difficulty. If electronic engineers would stick to the so-called radiometric system of units for specifying optoelectronic devices—watts/m^2, watts/steradian, etc.—the confusion would be very much reduced.

Radiometric and Photometric Equations and Units

Definition	Radiometric		Photometric	
	Name	Unit (SI)	Name	Unit (SI*)
Energy	radiant energy	joule	luminous energy	lumen-sec
Energy per unit time = power = flux	radiant flux	watt	luminous flux	lumen
Power input per unit area	irradiance	W/m²	illuminance	lm/m² lux
Power per unit area	radiant exitance	W/m²	luminous intensity	lm/m²
Power per unit solid angle	radiant intensity	W/steradian	luminous intensity	candela
Power per unit solid angle per unit projected	radiance	W/m² steradian	luminance	candela/m²

ILLUMINATION CONVERSION FACTORS				
1 lumen - 1/680 lightwatt (at 555 nm)		1 watt-second = 1 joule = 10^7 ergs		
1 lumen-hour = 60 lumen-minutes		1 phot = 1 lumen/cm²		
1 footcandle = 1 lumen/ft²		1 lux = 1 lumen/m²*		
Number of ⟶ Multiplied by ⟶ Equals number of	Footcandles	Lux*	Phots	Milliphots
Footcandles	1	0.0929	929	0.929
Lux*	10.76	1	10,000	10
Phots	0.00108	0.0001	1	0.001
Milliphots	1.076	0.1	1,000	1

LUMINANCE CONVERSION FACTORS						
1 nit = 1 candela/m²*						
1 stilb = 1 candela/cm²						
1 apostilb (international) = 0.1 millilambert = 1 blondel						
1 lambert = 1,000 millilamberts						
Number of ⟶ Multiplied by ⟶ Equals number of	Foot-lamberts	Candelas /m²	Milli-lamberts	Candelas /in.²	Candelas /ft.²	Stilbs
Footlamberts	1	0.2919	0.929	452	3.142	2,919
Candelas/m²*	3.426	1	3.183	1,550	10.76	10,000
Millilamberts	1.076	0.3142	1	487	3.382	3,142
Candelas/in.²	0.00221	0.000645	0.00205	1	0.00694	6.45
Candelas/ft.²	0.3183	0.0929	0.2957	144	1	929
Stilbs	0.00034	0.0001	0.00032	0.155	0.00108	1
*International System of Metric Units—recommended standard						

Figure 6-18. Conversion tables for photometric and radiometric quantities.

Stimulated emission

In an ordinary LED, when excited electrons fall to lower energy states and light pulses are emitted (Figure 6-10), the occurrence of drops from a higher energy state E_c to a lower state E_v are spontaneous and random events. The only definite statement that can be made about the occurrence of the events is that their rate is proportional to the number of electrons in the upper state. The individual light pulses that result from these random energy-level drops are thus not in phase and the emitted light is *noncoherent*.

However, if many excited electrons could be made to wait in a metastable state and then *stimulated* to drop simultaneously, the radiated light from these drops would be in phase. If the process could be arranged to synchronize itself over a period of time, the radiated energy during this interval would be in phase and *coherent*. Coherent radiation has a very narrow bandwidth, or single wavelength, which is called *monochromatic* radiation (Figure 6-19).

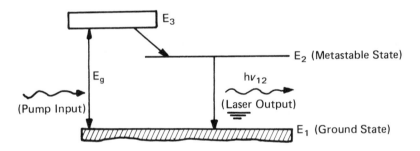

Figure 6-19. When electrons are pumped to a high energy level, E_3, and then fall to a state of relatively long lifetime as provided by chromium ions, E_2 (a metastable state), then large numbers of electrons are available for stimulated return to the ground state and the generation of a coherent light output.

Devices that behave in this manner are called LASERs. The acronym stands for *light amplification* by *stimulated emission* of *radiation*. The light from a LASER, depending upon its type, can be in a continuous beam or in short bursts of very high energy radiation, sometimes with peak powers in the megawatts.

There are many types of LASERs. Some use special crystals. The original LASER used a ruby crystal. Ruby is Al_2O_3 with about 0.05

percent chromium (Cr), and belongs to the sapphire family. When Al_2O_3 is doped with Ti, Fe or Mn, other variously colored sapphires are obtained; most of them can be artificially grown as single crystals.

Gases and liquids can also be made to *lase*. And the LED, when properly configured, can be operated as a LASER.

LEDs can operate as LASERs

To make a configuration lase, you must provide excited electrons at an elevated energy level and a means of retaining some of the generated radiation within the configuration to provide synchronization of the energy drops and an accumulation and build-up of energy before it is lost to the outside. Excitation in a ruby LASER is done with a high-intensity flash tube to produce optical excitation. This process is called *optical pumping*. Gas LASERs use high-voltage discharges to produce ionized-gas plasma; a LED LASER provides the excitation with its forward-bias current.

Light that is released into an *optically resonant cavity*, whose ends are partially reflective and whose length is an integral number of half-wave wavelengths, would allow several reflections between its ends of a portion of the light before the light leaves the cavity (Figure 6-20). The presence of such a buildup of energy in an optical cavity then in turn synchronizes further electron energy drops and the coherent generation of radiation.

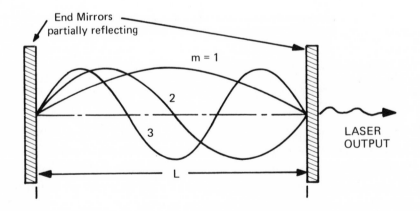

Figure 6-20. An optically resonant cavity is created by reflective, optically-parallel surfaces that are an integral number of half wavelengths apart—L = m λ/2.

The more reflective the optical cavity ends, the greater the energy retained in the cavity. This allows a greater buildup or amplification of the light. Some special LASERs use rotating mirrors and other means called *Q-spoilers* or *Q-switchers* to permit large buildups over periods of several microseconds; these can produce megawatt power pulses (Figure 6-21).

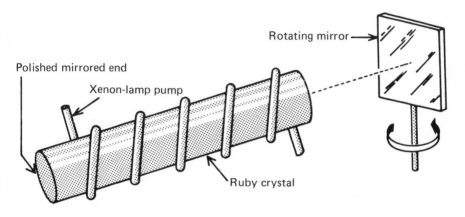

Figure 6-21. A Q-switched ruby LASER can use a rotating mirror as one reflecting face to permit large energy buildups within the crystal, before the light is permitted to radiate out.

In an LED, two opposite ends of the pn junction can be carefully polished and made optically parallel to create an optically resonant cavity (Figure 6-22). Such a configuration will lase when the bias current is raised above a threshold value and essentially monochromatic high-intensity radiation can be attained.

At low bias currents, an LED will emit spontaneously, like an ordinary LED, over a bandwidth that represents the difference between the top and bottom of the conduction and valence bands ($E_{ct} - E_{vb}$) and E_g (Figure 6-23(a)). As the current rises above the threshold, stimulated emission starts at frequencies that correspond to integral half-wavelengths of the cavity. At still higher currents, a single frequency or set of frequencies will dominate above a much lower background of spontaneous emission.

High currents generally mean that LED LASERs can't operate continuously at room temperature. They are thus often pulsed with current bursts of about 200 ns in the range of 1 to 5 A, and can deliver

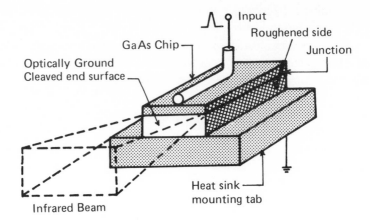

Figure 6-22. With optically ground end surfaces a diode LED, when properly forward biased, can be made to lase.

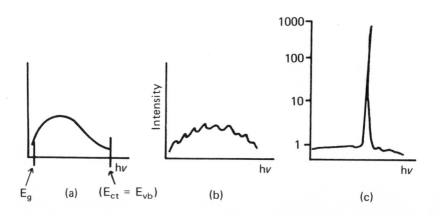

Figure 6-23. At low forward bias a LED LASER provides incoherent emission, like an ordinary LED (a). At the threshold bias the diode starts to lase (b) and at high bias currents a dominant frequency is generated (c). Note the logarithmic scale that describes the relative intensities between (a), (b) and (c).

about 0.1-W outputs. However, there are some low power LED LAS-ERs of limited life that operate continuously in the infrared region at room temperature when adequate heat sinking is supplied. Some nitrogen-cooled GaAs units operate in the near-infrared and visible red regions.

Optoelectronic devices come in pairs

When a photoemitter is matched to a photodetector device, the frequency-range and light-intensity compatibility between the devices tends to optimize performance and reduce sensitivity to external interfering sources. A number of applications for optically matched pairs are found in data handling, communications and industry. Optoelectronic pairs are used for sensing holes in cards and tapes, where they are arranged in arrays to detect a coded character, or in reflective configurations for picking up encoded patterns in reflective arrangements. Counting on conveyer lines and intrusion alarms for security control are other important applications.

There are many situations in which data are transmitted between two circuits that must be electrically isolated from each other. Such isolation was previously often provided by electromechanical relays, transformers or specially balanced circuits; now optically coupled pairs can handle most of these requirements. Optoisolators are usually combinations of infrared-emitting diodes and photosensitive diodes, transistors, SCRs or photoconductive cells that are spectrally matched to the emitter. The input is applied to the LED terminals and the generated light couples the input and output circuits. The light between LED and photosensor may be transmitted by fiber optics or a lens. Electrical isolation between the input and output is very high, typically in the neighborhood of 10^{11} ohms, and often units can separate kilovolts between input and output terminals.

Inputs frequently range around 1.5 V with currents from 10 to 100 mA; outputs can be microamperes to amperes with speeds into the megahertz. The output-input relationship is often characterized by an emitter-sensor current *ratio*. For example, an output of 25 mA and an input of 5 mA produces a transfer ratio of 25/5 = 5, but transfer ratios can be anywhere from 0.2 to 20. Most units have fixed ratios; some can be mechanically altered. The varieties are almost endless.

One large area of application is in the field of "solid-state relays." With optical isolation and power-level output circuits, these units provide a tempting alternative to bouncing, arcing and dirt-catching electromechanical units that wear out too quickly.

Chapter 7

A Practical Guide to Integrated Circuits

The integrated circuit (IC), or microcircuit, is an extension of the development of the transistor. Integrated circuits are made with materials and processes similar to transistors, but with ICs, complete electronic circuits are formed on a tiny piece of semiconductor material. A typical (small-scale) IC wafer is roughly 50 mils square and 10 mils thick and can contain over 50 separate electronic components. Medium-scale integrated circuits (MSI) have dozens of completed circuits (not merely components) such as gates, flip-flops and amplifiers; large-scale integrated circuits (LSI) can contain in the hundreds of circuits; thousands of circuits are also quite feasible.

On an IC, typical circuit components might need only 4×6 mils (1 mil = 0.001 in.) for a transistor, 3×4 mils for a diode and 2×12 mils for a resistor. Components are packed together on the surface of the semiconductor wafer and interconnected by a metal pattern that is evaporated onto the top surface. Leads are attached and the wafer, often called a *chip*, is then sealed and packaged in a variety of forms.

Purity makes it all possible

Both ICs and transistors trace their beginnings to the early 1950s when an important breakthrough produced extremely pure single-crystal semiconductor materials such as silicon. Purities to one part in 10^{10} were attained. As discussed in Chapter 2, such extreme purity is needed to allow manufacturers to accurately reproduce device characteristics and to obtain economic yields.

Silicon is today's most versatile and popular semiconductor mate-

155

rial, and we will describe the process of its purification in some detail. However, the steps that describe the purification of silicon, with appropriate variations, also apply to other materials. The first step in the purification of silicon is to reduce silica (silicon dioxide as in quartz beach sand) by heating it with carbon (coke) in an electric furnace. This produces silicon that is about 98 percent pure. Additional chemical processing, which includes forming silicon tetrachloride, repeated distillation and then reduction again with hydrogen, further purifies the silicon.

The polycrystalline silicon that is thus obtained must now be converted into a single crystal, and in the process still more purification is attained. This process is referred to as *crystal pulling*. The polycrystalline material is placed inside a pure quartz crucible that is enveloped in an atmosphere of the inert gas, argon. The quartz crucible is surrounded by a graphite susceptor that is capable of being heated by rf induction. Graphite is used because it is a conductor, and quartz is an insulator (Figures 7-1(a) and (b)).

The silicon in the crucible melts and its temperature is adjusted to barely above the melting point. At this point, a seed crystal, which is a small, single-crystal piece of silicon, is lowered into the molten material. The cooler seed crystal causes the molten silicon that it touches to cool and solidify about itself, and the solidifying atoms arrange them-

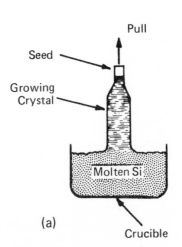

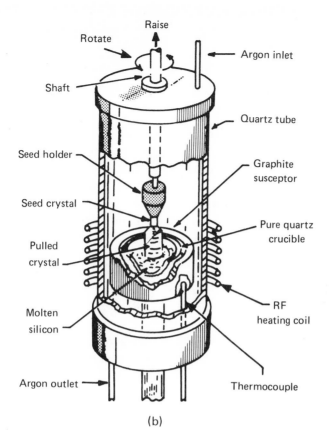

(b)

Figure 7-1. The crystal-pulling method (a) and (b) requires a crucible, which can contaminate the semiconductor material. The floating-zone method (c) uses no crucible and avoids this potential source of trouble.

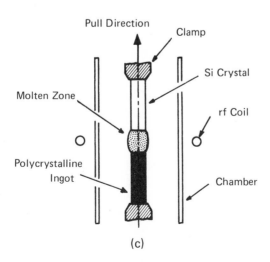

(c)

selves to add to and enlarge the original seed crystal. The seed crystal is rotated to provide uniform exposure (about 60 rpm) and it is slowly raised at about 1 in/h.

A typical pulled silicon crystal is about 1 to 2 inches in diameter and about 12 inches long. A p- or n-type dopant can be added to the initial melt to provide the crystal with required initial properties.

The presence of the crucible can be a source of contamination, since the silicon dioxide of quartz can release oxygen. To eliminate contact of the molten material with any container, semiconductor crystals are often grown by the *floating-zone* method (Figure 7-1(c)). This technique uses a polycrystalline ingot that is held vertically at its ends, while a small zone of the material is heated to the melting point. In this way the molten material is never in contact with a container and contamination can be more easily avoided.

A seed crystal that is included at the starting end of the ingot forms the nucleus for single-crystal growth as the molten zone is moved slowly along the ingot. Localized rf induction heating is a common method for concentrating the heat into a narrow zone. Several passes of the molten zone along the crystal can provide considerable purification of the crystal. Impurities are concentrated into the end-zones of the boule (crystal) and can be cut off.

The process of recrystallization is a well-known method for purification of any chemical. When a crystal forms, impurities that do not fit the crystal structure of the bulk of the material tend to be thrown off. An example is salty sea water: when frozen, the resulting ice is practically salt free.

Growing thin-film, single-crystal layers

Many semiconductor devices are fabricated from thin-film, single-crystal layers that are grown on top of another crystal called a *substrate*. The process used for this film growth is called *epitaxial growth*. Substrate crystals are usually wafers sliced from single-crystal ingots made by such methods as the previously described crystal-pulling or floating-zone techniques. The substrate may be of the same or similar lattice-structured material. In the process, the substrate serves as the crystal seed.

An epitaxial layer can be grown by dipping the wafer into a molten solution of the desired material and then withdrawing it after the solution has been cooled and the growth has taken place on the substrate.

Another method uses volatile silicon tetrachloride. Hydrogen gas is bubbled through the tetrachloride and the vapor plus the hydrogen is passed into a reaction chamber where the substrate slice is heated to 1200°C. The silicon tetrachloride dissociates and silicon is deposited onto the surface of the heated substrate to form the epitaxial (*vapor epitaxy*) layer. Growth rate is about 0.04 mils/min. Dopants can be added to the hydrogen stream to provide controlled p- or n-type layers.

Dopants also can be introduced into pure semiconductor material by a process called *diffusion*. The wafer is heated to between about 800° and 1250° C in the presence of a controlled density of impurity atoms, but this process is very slow. The penetration rate for the dopant is about 0.1 mil/h.

Often a two-step method is used. First the wafer is exposed to high-concentration dopant vapor to form a high-concentration surface layer. This step is called *deposition*. The slice is then heated to a higher temperature in a diffusion step to allow the dopant to spread through the whole layer.

Separate dopants can be diffused, one after the other, to form alternate p and n junction layers. A subsequent diffusion concentration must be greater than the preceding one to reverse the layer's characteristics from the previous one. This alternating technique is often used in manufacturing *monolithic* ICs.

Categories of integrated circuits

ICs can be categorized by their method of fabrication or use. The most common are *monolithic* or *hybrid* and *linear* or *digital*. Operational amplifiers and most other analog circuits are linear, and gates, flip-flops and other on-off circuits are digital.

ICs that are included entirely upon a single chip of semiconductor material are monolithic. Any additions to the semiconductor chip such as metallized patterns, resistors, capacitors or insulation areas are securely bonded to the chip's surface. A hybrid, on the other hand, may consist of several interconnected monolithic elements, often in combinations with individual discrete components such as transistors, resistors or capacitors. The hybrid can also be made of all discrete parts.

In a monolithic IC, all circuit components, both active and passive, are formed at the same time. Since all components are contained on a small rigid chip, monolithics can be batch fabricated: hundreds or even thousands of identical circuits can be built at the same time at low cost. Hybrids, however, provide greater circuit flexibility, better isolation against component interaction and greater component range and precision, especially in the choice of resistors and capacitors, and they can handle higher power.

Hybrid: a construction technique

Hybrid technology combines the techniques of monolithic and discrete, and uses *thick* and *thin film* components for obtaining the "best" solution to a particular design problem. "Best" implies, of course, a compromise between cost and technical perfection. One of the drawbacks of a straight monolithic approach to many designs is its temperature sensitivity and the restricted range of values of passive components that can be manufactured by this method. Many applications that require close tolerances, high temperature stability, higher power handling ability and wide band widths, etc. cannot be met by the monolithic technique. This is especially true for linear or analog circuits. Digital ICs, however, are most often straight monolithic.

Active components (transistors, diodes, FETs, etc.) for hybrid packaging of circuits are formed as for monolithic.

Passive components (resistors, capacitors and rarely, inductors) are made by both *thick* and *thin-film* techniques. Film techniques are assemblies of components that have been formed on and are affixed to an insulating base material (substrate).

Thin films have a thickness between 0.001 and 0.1 mil and are usually deposited on a substrate wafer of silicon dioxide by evaporation or sputtering methods. Substrate sizes normally range from 0.25 to 2.0 inch square.

Resistors, in values from low ohms to megohms, are made by evaporating nichrome or tantalum to form strips between terminal regions of high conductivity. The values are controlled by varying the length, width and film thickness, and by changing the material composition.

Thin film capacitors form when conducting areas are deposited on both sides of a thin film of dielectric material. The thickness and material of the dielectric and the conducting area dimensions control the value of capacitance. Materials such as tantalum oxide, aluminum oxide and silicon dioxide are used for the dielectric.

There are many special processes and trade secrets that go into the making of thin film components; the state of the art is under continuous change and development. A great many manufacturers, however, use tantalum for all resistor and capacitor formations. A tantalum film is sputtered over the substrate and the pattern of resistors and capacitors is photoetched onto this film. The film is then oxidized to a desired depth to form dielectric for capacitors. The top electrode for the capacitors and the conducting pattern to interconnect the resistors and capacitors is then deposited using gold or platinum.

Thick films are normally applied to a supporting substrate by the silk-screen process. A popular substrate is the ceramic alumina. Substrate size, on the average, is about 0.5 inch square and 60 mils thick. The first step in the fabrication process is the application of a metallized ink interconnection pattern by the silk-screen process, followed by firing the assembly at a temperature of 700°C. *Resistors* are formed by using a metal-glass slurry, applied through another silk screen and then fired at 700° C. *Capacitors* are generally discrete miniature components that are soldered or welded into the circuit, especially where high value capacitances are required. However, film capacitors can be fabricated directly on the substrate at low capacitance values.

Thick film capacitors often use a platinum bottom electrode deposited directly onto the ceramic substrate. The dielectric is a paste made of ceramic and glass that is screened over the bottom electrode. After firing, a top electrode of platinum is applied. Finally, the interconnecting pattern that was applied in the first step is coated with solder to a thickness of 2 to 3 mils.

The *passive components* of both thick and thin film processes can be trimmed to precise tolerances. This is a major advantage of the hybrid design. Recent developments in the functional, or dynamic, trimming of resistors by LASER techniques, to tolerances of better than 0.05 percent in automatic modes and 0.001 percent manually can tailor a group of components to a particular functional value *after the circuit has been assembled*. Also, thick-film resistors can be trimmed to better than 1 percent by an abrasive-air removal process.

Older hybrid methods used encapsulated, discrete, active

semiconductor components that were soldered into the circuit, but it is less expensive and more common to use active components on chips. These are attached to mating conducting pads by a variety of methods that include thermocompression bonds, epoxies, welding and soldering.

The hybrid design can encompass a large variety of processes. There are as many different ways of doing hybrid manufacturing as there are practitioners of the art.

Packaging the hybrid

The first step in selecting a hybrid package is counting the pins. The pin count may have more impact on package selection than the amount of circuitry. The next step can be defined only after several questions are answered:

- What size substrate is needed?
- Is the circuit so complex that it requires more than one package?
- Can the circuit be divided into individually testable functions?
- What style package should be used?
- Are there special mechanical and environmental requirements?

Table 7-1 lists some standard package styles with their pin capacities, available substrate areas, power dissipation capabilities and thermal resistance.

You can estimate the substrate area required in the following way:

Assume one unit of area is a resistor of area 0.015 in.^{2} Assign a number of units to each component per Table 7-2 and total the number of units for the circuit. Determine the substrate area required as follows:

$$A_S = (0.015 \text{ in.}^2)(U_T)$$

where A_S = required substrate area, and U_T = total units. A package choice based on available substrate area can then be chosen from Table 7-1.

The unit system is strictly a short-cut method; thus a considerable amount of judgment and common sense must be applied. The following general comments should be considered:

Table 7-1. Some popular IC package styles.

PACKAGE TYPE	METHOD OF SEAL	NUMBER OF TERMINALS	AVAILABLE SUBSTRATE AREA (sq. in.)	Θ (C/W)	+25°C POWER RATING (W)	OUTLINE DRAWING
Metal Pin Packages	Cold Weld	10 20 25 30	0.465 0.850 0.518 1.250	60 30 40 20	2 4 3 6.5	A
Metal Ribbon Lead Packages	Stitch Weld	12 22	0.340 0.465	90 55	1.4 2.3	B
Dual In-Line Packages	Epoxy	14 16	0.086 0.098	80 80	1.5 1.5	C
Ceramic Plug-In Package	Epoxy	40	1.275	15	8	D

Courtesy of Beckman Instruments.

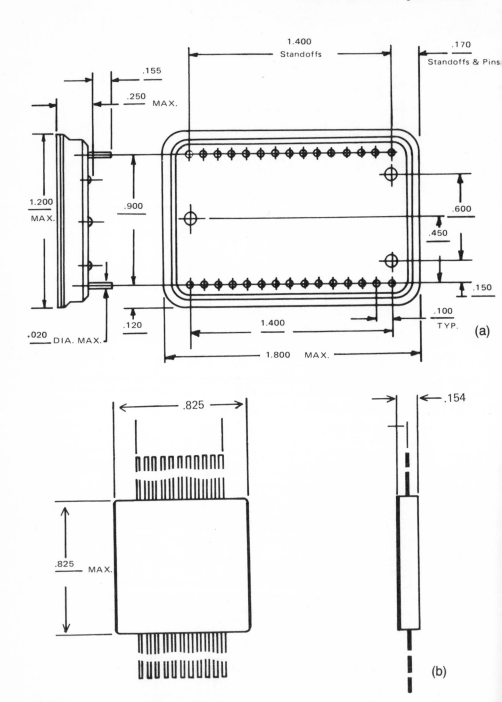

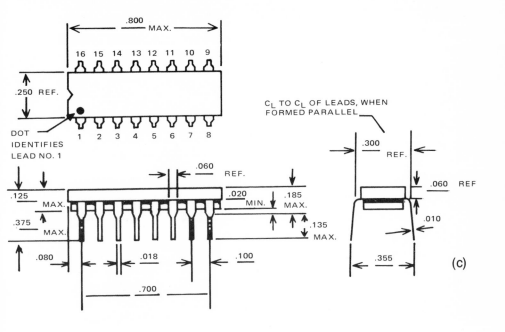

(c)

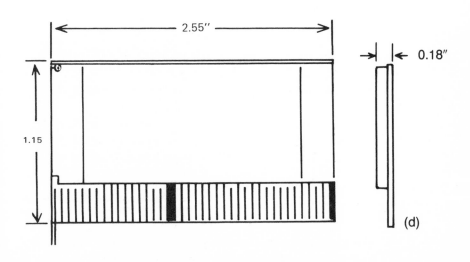

(d)

COMPONENT TYPE	UNITS PER COMPONENT
Resistors (Cermet, Thick Film) General Purpose (up to 100 mW).............1.0 Units	
Capacitors Screened (Cermet) ....270 pf/Unit Chip Capacitors 0.1″ × 0.1″ ....2.0 Units	
Diodes, Passivated Chip Signal/Switching ....0.5 Units Zener/Reference ....0.5 Units Schottky/Hot Carrier ....0.5 Units	
Transistors, Passivated Chip Bipolar Small Signal ....0.5 Units Bipolar Low/Medium Power ....1.0 Units JFET ....0.5 Units	
Integrated Circuits, Passivated Linear ....2.0 Units Digital ....4.0 Units MOS Arrays ....0.5 Units/Lead MSI Devices ....0.5 Units/Lead	

Courtesy of Beckman Instruments.

Table 7-2. Component Unit Areas

- It may be easier to deduct the area required for large component chips before applying the unit system.
- Some circuits naturally flow from input to output; the circuit layout should take advantage of this.
- Packaging efficiency is usually greater for large packages.
- The component density of 0.015 in.^{2}/unit is a moderate density level and can be readily achieved provided pin assignments are not fixed before the layout is designed.

The easiest solution to the choice of a package is a standard design that is compatible with system constraints. This avoids a development program and has the following advantages:

- The package design, assembly and processing details are well defined and, hopefully, already debugged.

- Tooling charges are minimized.
- The delivery cycle is short.
- Price can benefit from the standard package's volume.

Temperature is another constraint

Along with substrate area, allowable temperature rise presents another important packaging constraint. The package temperature rise (T_R) is a function of the total power dissipation (P_T) of all internal circuit elements as follows:

$$T_R = T_C - T_A = (P_T)(\Theta_{CA}),$$

where T_C = case temperature,

T_A = ambient temperature

and Θ_{CA} = case-to-ambient thermal resistance.

An approximate value of Θ_{CA} for a package in free air and minimum pin conduction is 35° C rise per watt of power dissipation per square inch of package area (35° C/W/in.²). For example, a circuit that dissipates 1 W rises approximately 35° C above ambient in a one-inch square package, or 70° C above ambient in a one-half square-inch package. This general rule is conservative and a very safe first approximation for most printed circuit board applications.

Although the case temperature might be safe, individual component temperatures must also not exceed their safe limits. The maximum allowable junction temperature for silicon devices depends on the application and in some cases is limited by system specifications. The general rule is to never allow a junction to exceed +200° C in general utility applications. High reliability applications often further limit temperature to +150° C.

Monolithics are small

In *monolithic ICs*, all circuit elements, both active and passive, are formed simultaneously on a single *wafer* or silicon. The elements are interconnected by metalized strips deposited on an oxidized surface of the silicon wafer. The same circuit can be repeated a large number of times on the individual wafers of a slice of semiconductor. For instance, a single slice 1.5 inches in diameter will contain upward of 500 wafers of a 50-mil square IC (Figure 7-2).

Even though monolithic components are much less precise than discrete ones and involve more compromises, ICs do have highly de-

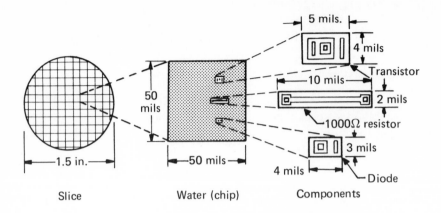

Figure 7-2. A slice can be cut into as many as 500 wafers, and a wafer can easily support 50 components in monolithic ICs.

sirable properties (Table 7-3). The most important such property is that two like elements in close proximity tend to have very nearly identical characteristics. This property can be exploited by making circuit performance depend on the ratios of elements rather than on absolute values. Ratios of monolithic elements are more precise and predictable than are discrete components.

Another advantage is the uniformity of the temperature over the chip. All elements are within a short distance of each other. Since they are part of a single chip of fairly high thermal conductivity, they are likely to have the same temperature. If the temperature of the chip changes, the ratios of the element values tend to stay the same. Because of this thermal tracking, ICs can be more stable than discrete circuits.

A breadboard design made with discrete circuit elements that can also be produced with the diffusion process is the first step in a monolithic's creation. The circuit elements are then converted to geometric patterns enlarged 500 times. A series of photomasks that contain repetitions of the enlarged circuit layout is photographically reduced to actual size and in several steps of exposure, etching and diffusion recreate the circuit in IC form (Figure 7-3).

All *diffusions* are done under a passivating layer of silicon dioxide, except for small cutouts through which the dopants are introduced. This protective method assures good yields of the semiconduc-

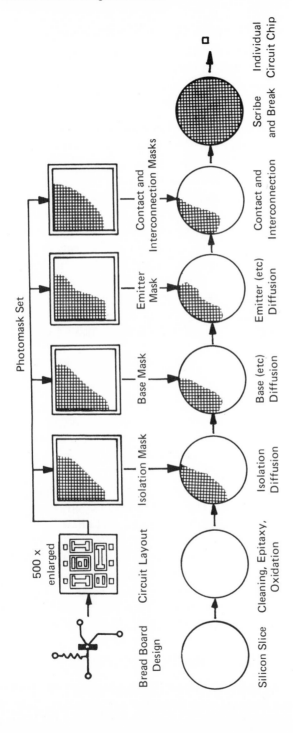

Figure 7-3. The patterns of a series of photomasks are sequentially exposed with ultraviolet onto repeatedly applied photoresist material. The slice is selectively etched after each exposure and the required dopants diffused into the layered semiconductor surfaces.

tor devices. Figure 7-4 illustrates the combination photolithographic, etching, diffusion, heating, vaporizing and depositing processes that go into the making of a semiconductor-device chip.

Component must be isolated

Note in Figure 7-4 that an n-type epitaxial layer is first grown on a p-type substrate. The surface of the epitaxial layer is oxidized and the oxide selectively removed. A p-type diffusion then forms p-type regions that extend through the epitaxial layer to the p substrate. This leaves the n-type regions each separated from the substrate by a pn junction. When the IC circuit is operated, the pn junctions are all biased in the reverse direction. The p substrate is made more negative than any part of the circuit. Each junction then presents a very high resistance, which isolates the n-type "islands." This explains the use of the name "isolation-mask pattern."

Other methods are also used in ICs to isolate a circuit's components. With oxide isolation, a layer of silicon oxide is formed around each element. A slice of n-type single-crystal silicon has channels etched in the surface between the locations planned for each element. Next the surface of the slice, including the channels, is oxidized to form a continuous layer of insulating silicon dioxide and polycrystalline silicon is deposited on top of the oxide. Finally the slice is inverted and the original silicon is lapped down so that only the regions between the channels are left. Each of these is a region of single-crystal silicon isolated by the layer of silicon dioxide and supported on the substrate of polycrystalline silicon.

Another system of element isolation used for special applications is called *beam-lead isolation*. The circuit elements are formed in a wafer of silicon in the regular manner. The metallization used for interconnection is thicker than that normally used, and the silicon between each element is completely removed by etching from the back side. The etchant does not attack the metallization, and the result is that each element is completely separate, supported from the top by the metallic connections. A thermosetting plastic can be used to fill the spaces between the elements to restore mechanical support.

Photolithography controls the accuracy

For economy and speed of operation, integrated circuits should be small. Although the ultimate size limitation is dictated by heat, the

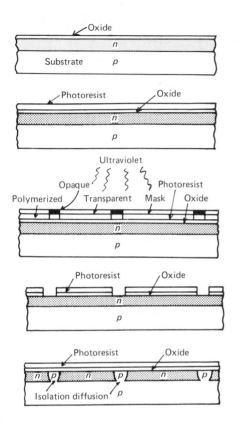

1. Epitaxial growth of n on p-material (or vice versa) base, cut from single-crystal 1 to 2-inch diameter and several mils thick. Epitaxial is then covered with silicon dioxide by heating to 1000 C in a flow of oxygen.

2. Photoresist is applied.

3. Isolation pattern is exposed.

4. Pattern is etched into oxide.

5. Diffusion, reoxidation and reapplication of photoresist. Isolation pattern is attained. Islands of n-type material will surround components made in subsequent steps.

6. Expose through base-diffusion pattern and repeat steps 4 and 5.

7. Expose through emitter-diffusion pattern and repeat steps 4 and 5.

8. Similarly for the collector-diffusion and contact/interconnect patterns.

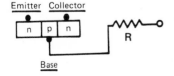

9. An npn transistor and resistor have been produced.

Figure 7-4. The steps from slice to electronic parts.

practical limit results from the accuracy of the dimensions of the circuit elements, which are determined by the photolithographic process.

A series of master drawings of the circuit is made on a greatly enlarged scale and each drawing represents one stage in the process. Each master drawing is then photographically reduced and reproduced many times to form a matrix on a film transparency called a mask, in actual size. The masks contain patterns for all the circuits to be made on the slice.

Since the masks are used sequentially during the processing of the slice, the pattern of each successive mask must align precisely with the previous patterns. *Alignment* is extremely important.

Resolution is defined as the number of lines per millimeter that can be placed on the mask when the space between the lines is equal to the line width. Because of distortion in the lenses used to reduce the master drawing to mask size, mask resolution tends to fall off toward the edges of the circuit. This effect is more pronounced with large circuits. When the mask is illuminated, there is scattering of light in the photoresist and diffraction around the edges of the mask which tends to blur the image. Finally, because the silicon dioxide cannot be etched away with perfect uniformity, the oxide edge does not exactly follow the pattern outlined by the exposed resist.

Fabricated components are therefore not absolutely accurate. The parameters that are influenced by geometry are subject to variation, but for a given process, the variations are generally constant so that small devices have greater percentage errors than large ones. This is the practical limit to small size.

Fabricating the transistor and diode

In ICs the bipolar transistor is probably the most important. The process used to fabricate silicon ICs was originally developed for discrete transistors (called the planar process), so it is not surprising that the integration of transistors requires fewer compromises than other components. An integrated transistor performs almost the same as a discrete transistor.

There are differences, however: in IC transistors the contact to the collector region is made through the top surface, not through the substrate (Figure 7-4), and the substrate must be electrically isolated from the collector region. The current path between the collector contact and the collector-base junction is via a narrow region of high-resistivity n-type material; therefore, the collector output resistance of integrated bipolar transistors is not negligible.

To reduce collector resistance, an n-type region of high dopant concentration (n+) is first diffused under the normal n material. This buried layer short-circuits the higher resistance of the n-material path.

The fact that an integrated transistor is isolated from the other components by a pn junction causes effects not found in discrete transistors. The junction between the collector and the substrate, which is reverse-biased to isolate the device, has capacitance. This degrades the performance of the transistor at higher frequencies. Also, leakage current between the collector and the substrate can become significant in low-current applications.

In the IC transistor, the isolation junction creates a parasitic transistor. The top three layers of the structure form a desired npn transistor, but the bottom three layers form a parasitic pnp transistor. When the collector and the substrate are negative with respect to the base of the npn transistor, the parasitic transistor is in its active region and can cause unexpected problems that are difficult to diagnose.

Because of their importance, resistivities and doping profiles of the various layers are chosen to optimize transistors. Generally one or more of the layers or regions provided for transistors are also used for diodes. In fact, integrated diodes can be regarded as transistors whose terminals have been connected to give the desired diode characteristic. Several diode configurations are possible; each has certain advantages.

IC diodes are therefore prepared by forming pn junctions at the same time as the transistor junctions. Figure 7-5(a) shows a diode in which the cathode is the same n-type collector region as in a transistor. The p-type anode is formed when the transistor bases are diffused. This type of diode has the same reverse-voltage capability as a transistor base-collector junction and is widely used for general purpose circuits. Where fast switching speeds are required, base-emitter diodes are used Figure 7-5(b)). This technique provides a low-voltage diode. To avoid unwanted effects because of transistor action the base p-type anode contact is arranged to short-circuit to the n-type collector region.

Several other configurations are also possible, but they offer little advantage over these two.

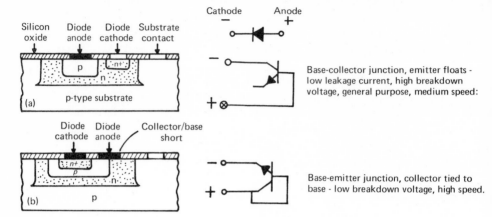

Figure 7-5. IC diodes are usually based upon transistor configurations. Several other configurations are also possible.

Resistors are formed with transistors

Diffused resistors are also formed at the same time as transistors. No additional processing steps are necessary to do this; however, it presents a serious problem. Diffused resistors exhibit a large change of resistance with temperature, and the resistance range is limited. Deposited film resistors (as in hybrids), on the other hand, require additional processing steps but offer a much wider range of resistance, lower temperature coefficient of resistance and lower stray capacitance.

In either case, a layer of resistive material is characterized in terms of its sheet resistance. No matter how large or small a resistor square is, its resistance is determined only by the resistivity of the material and the thickness of the layer. To obtain a 1000Ω resistor from 100Ω/square material, the designer merely places ten squares in series and makes contact to the extreme ends. To save space, resistors often must assume odd shapes to fit the available area.

In an IC, several layers can be used for resistors. The one usually favored is the p-type layer created during the diffusion of transistor bases (Figure 7-4). Its sheet resistance is approximately 100 to 200Ω/square—a compromise between maximum resistivity and low temperature coefficient.

Large voltage differences between the p-type resistor and the n-type epitaxial layer must be avoided because a pinch-off effect similar to that in FETs can produce a nonlinear resistance characteristic.

The two other layers in silicon ICs are less suited for resistors. The highly doped n+ emitter-diffusion layer has a sheet resistivity of only a few ohms per square—too low for most resistor values in ICs, but sometimes useful. The n-type epitaxial layer, however, has a sheet resistance which approximates that of the base region. Satisfactory tolerance is difficult to obtain and proximity to the substrate results in large stray capacitance.

For many reasons, it is difficult to reproduce diffused resistors to better than ±10 percent; however, the ratio between two resistors formed side by side can be reproduced very accurately to within ±1 percent. Circuit design for integrated circuits thus tends to use resistance ratios as a controlling factor, rather than absolute resistance values. Improvements in processing will no doubt gradually allow better reproducibility of diffusion-surface components.

Capacitors are needed too

Two basic types of capacitors are available to the IC (Figure 7-6). Each consists of two, low-resistance layers separated by a carrier-free region. In a *thin-film integrated capacitor*, one of the layers is deposited metal; the carrier-free region is silicon dioxide, a dielectric mate-

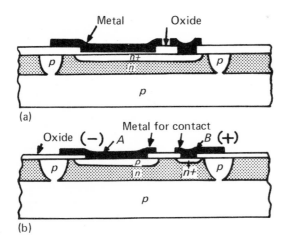

Figure 7-6. Two capacitor types that are most used in monolythic ICs are a thin-film type (a), which uses the SiO₂ layer as the dielectric, and a pn-junction type (b), which is a reverse-biased diode that behaves as a capacitor.

Component	Typical value	Tolerance	Temperature coefficient
Transistors and Diodes			
Current-amplification factor	50	+ 50 −30%	+ 0.5%/° C
Matching β between (transistors in close proximity)	—	± 10%	± 0.05%/° C
Base emitter diode forward voltage drop (low current)	0.65 V	± 3%	−2 mV/° C
Matching V_{BE} between ·(transistors in close proximity)	—	± 2 mV	± 10 μV/° C
Base-emitter diode reverse breakdown voltage	7 V	± 5%	+ 3 mV/° C
Collector-base breakdown voltage	>45 V	± 30%	—
Collector-substrate breakdwon voltage	>60 V	± 25%	—
Resistors			
Resistance of diffused resistors (base layer)	150Ω/sq	± 25%	+0.2 %/° C
Resistance of deposited resistors	50-100Ω/sq	± 8%	0.01%/° C
Matching resistors in close proximity	—	± 3%	±0.0005%/° C
Capacitors			
Capacitance of diffused capacitors	—	± 25%	—
Capacitance of deposited capacitors	—	± 20%	—
Matching capacitors in close proximity	—	± 3%	—
Junction FET's (depletion mode)			
Pinch-off voltage	—	± 30%	—
Pinch-off current	—	± 30%	−0.5%/° C
Transconductance	—	± 50%	−0.2%/° C
Insulated-gate FET's (enhancement mode)			
Threshold voltage	—	± 50%	—
Transconductance	—	± 50%	± 0.1%/° C

Table 7-3. Monolithic component values and tolerances

rial. In *junction capacitors*, both layers are diffused low-resistance semiconductor material of opposite dopant type; the carrier-free region results from charge depleted at their pn junction.

The advantage of the thin-film capacitor is its simplicity. Capacitance is adjusted by etching the surface oxide to a thickness that will give maximum capacitance per unit area and still withstand the voltage. Because one of the plates is a semiconductor, the capacitance is somewhat voltage dependent.

An n+ region is diffused into the silicon at the same time as the transistor-emitter diffusion to form the bottom electrode of the capacitor, and the controlled thickness of silicon dioxide is formed on the surface of this region to produce the dielectric. The top electrode of metal is deposited at the same time as the interconnections.

The *integrated junction* capacitor is formed by diffusing p-type dopant into an isolated island of the n-type epitaxial layer. Usually this is done at the same time as transistor-base diffusion.

The junction capacitor is a diode (see Chapter 8) and can only be used with the diode reverse-biased. The value of capacitance per unit area is quite low; the maximum value is limited by economic considerations to about 100 pF. The capacitance of a junction capacitor is highly voltage dependent.

Against these limitations, IC capacitors have the advantage that they can be formed at the same time as the other elements with no additional processes.

A thin-film capacitor can also be built entirely on top of a chip as in a hybrid. The interconnection film is extended to form the bottom plate and the dielectric is applied over it; then another thin metallic film forms the top plate. This method uses chip area more efficiently, since the capacitor is built over previously diffused and protected elements. But such capacitors require additional processing steps which can increase cost and decrease yield.

Inductors are seldom used in ICs

Though inductors are seldom used in ICs, it is possible to make the coils of an inductor by successive deposition of conductive patterns and insulators. The "Qs" of such devices, however, are too low to be effective for most applications. Inductors can be replaced by piezo or mechanical transducers. One can also simulate inductance with active circuits via feedback arrangements. Since solid-state amplifiers are small and inexpensive, the simulation of inductances with active circuits is quite practical and many excellent circuits have been devised.

MOS ICs are smaller

So far we have discussed mostly the bipolar transistor and associated components, but IC MOSFETs have important advantages:

- They occupy only a fraction of the chip area required for bipolars (Figure 7-7). The average IC MOSFET has an area of 1 to 2 mils^2 versus 25 to 50 mils^2 for a bipolar equivalent.
- This space saving results mainly because the MOSFET, or MOS in brief, needs no separate isolation pads. As long as the source, gate and drain are operated above the substrate voltage (for n channel) the FET is electrically insulated. The elimination of the need for isolation is in itself an advantage, since several states in the fabrication steps are also eliminated.
- An MOS also behaves as a voltage-controlled resistor; therefore, by suitable biasing, excellent resistors of high values can be obtained in small spaces. Values as high as 100 kΩ can be placed in an area of 1 mil^2. An equivalent diffused resistor would take up over 150 mil^2.
- The gain of a MOS is a direct function of the device's geometry and thus determined mainly during design. In the bipolar, however, the gain is very strongly determined during processing in the emitter diffusion step.
- The MOS has a very high input resistance, typically $10^{18}\Omega$.
- The output of an MOS is bilateral (conducts equally well in both directions), which makes the drain and source interchangeable; thus there is no offset voltage to contend with and the output behaves as if it were a voltage-controlled resistor. This provides the circuit designer with a unique degree of flexibility not present in many other devices.

PMOS, NMOS and CMOS ICs

The first commercial MOS ICs were of the p-channel enhancement type (PMOS). The PMOS process was the easiest to control.

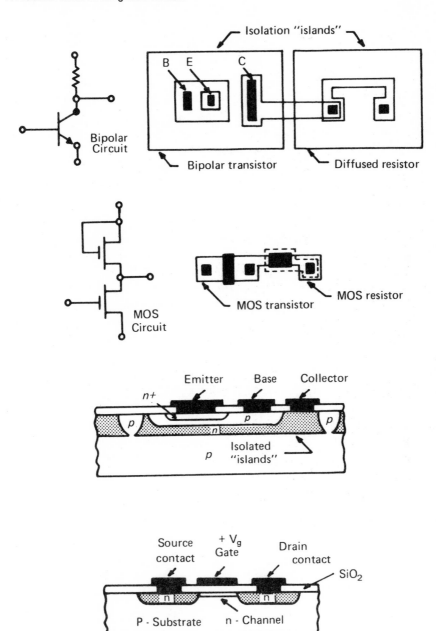

Figure 7-7. MOS construction takes less area than bipolar mainly because MOS does not need isolation "islands."

Advances in processes and materials produced n-channel enhancement (NMOS) structures and put NMOS and PMOS structures in the same circuit to form complementary (CMOS) devices. NMOS circuits are two to three times faster than PMOS circuits because electrons are more mobile charge carriers. The complementary, or CMOS, devices can very well become the most important of all. With both channel types combined in one circuit it is possible to gain performance advantages over all other varieties of ICs. A basic CMOS circuit consists of one p-channel and one n-channel transistor connected in series (Figure 7-8). The circuit is a highly effective digital switch with low power consumption in both the ON and OFF state. By interconnecting a number of such basic stages, a large variety of useful circuits with extremely low power consumption can be built. For example, a 16-stage binary counter consumes about 3 μW at 5 V. This is about 100,000 times less than comparable bipolar circuits. CMOS circuits are thus preferred for applications where the power is limited, and every battery-powered device is a candidate for their use. A CMOS pair can also be connected in parallel to handle both digital and analog signals in either direction.

CMOS circuits are also highly immune to spurious noise. This makes them particularly suitable for environments such as automobile engines. Circuit designers find that CMOS circuits can provide almost every switching and logic function.

An important market for CMOS circuits is in watches and clocks. Electronic timepieces have an accuracy impossible to attain by mechanical means. The circuit uses a quartz-crystal oscillator whose highly accurate fundamental frequency is divided down by counting circuitry to drive clock hands, LED or liquid-crystal displays.

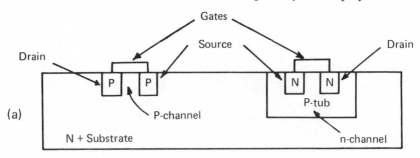

Figure 7-8. A p-channel, PMOS, FET (a).

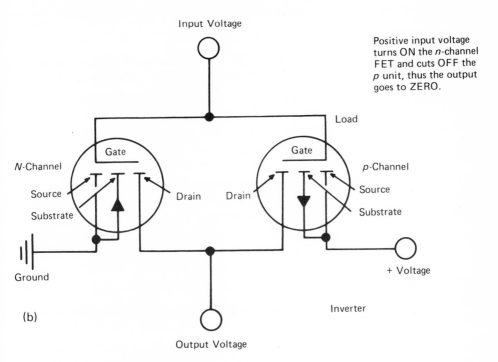

Positive input voltage turns ON the *n*-channel FET and cuts OFF the *p* unit, thus the output goes to ZERO.

(b)

Figure 7-8 (cont.). When placed in series with a p-channel, PMOS, FET, an n-channel, NMOS, FET forms a CMOS inverter (b).

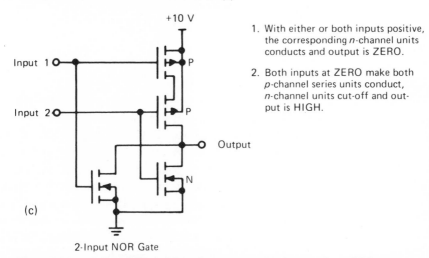

1. With either or both inputs positive, the corresponding *n*-channel units conducts and output is ZERO.

2. Both inputs at ZERO make both *p*-channel series units conduct, *n*-channel units cut-off and output is HIGH.

Figure 7-8 (cont.). Four FETs in (c) are interconnected to form a 2-input NOR gate. Almost any logic function can be made with CMOS.

From the system designer's point of view, one advantage of the CMOS technology is that the supply voltage may vary from a low of just over the combined threshold voltages of the series N and P channel FETs, about 2.2 V, to a high voltage level of just under the breakdown voltage of the devices, generally over 24 V, but limited for reliability to about 18 V. The advantage of a wide range single operating voltage is readily apparent.

Related to this wide power-supply range is a noise immunity. Noise immunity runs about 45 percent of the supply voltage so that with 16-V supplies, a noise immunity of 6.9 V can be expected. This is higher than that of other logic families.

Also inherent in CMOS devices is an extremely large OFF impedance—$2 \times 10^9 \ \Omega$ on a typical device. Since one gate of the complementary devices is always OFF, except during actual switching, currents are in pico-amps and quiescent power drains in nanowatts. This miserly use of power is perhaps the biggest single feature of CMOS that attracts the most attention.

Low power also means low heat dissipation and smaller power supplies. In a typical design the cost of the power supply and mechanical cooling can add up to as much as the ICs. CMOS reduces or eliminates the need for heat sinks, fans and large power supplies.

The one area in which CMOS, and MOS in general, don't quite measure up is in frequency response. Present metal-gate fabrication techniques have device capacitances which limit practical speeds to the 5-to-10-MHz range.

Connecting the chip to the outside world

The individual IC chip must be connected properly to outside leads and packaged conveniently for use in a larger system. Since the devices are handled individually once they are separated from the wafer, it is desirable to reduce the steps required in bonding individual leads from contact pads on the circuit to terminals of the IC package.

Early methods for making contacts from monolithic chip to package used fine gold wires. Later techniques used aluminum wires. The chip is first mounted solidly on a header or a metallized region of an insulating substrate. A thin layer of gold, perhaps combined with germanium or another element to improve the metallurgy of the bond, is placed between the bottom of the chip and the substrate; then heat and a slight scrubbing motion are applied to form an alloyed bond which holds the chip firmly to the substrate. This process is called die bonding. (Die is the singular form of dice; the term is used interchangeably with chip.) The interconnecting wires are then attached from various IC contact pads to posts on the header.

In gold-wire bonding, a spool of fine wire about 0.0007 to 0.002 inch diameter is mounted in a lead bonder and the wire is fed through a glass or tungsten-carbide capillary to a hydrogen gas flame jet. A ball forms on the end of the wire. In thermocompression bonding the chip, or in some cases the capillary, is heated to about 360° C, and the capillary is brought down over the contact pad. When pressure is exerted by the capillary on the ball, a bond is formed between the gold ball and the aluminum pad; then the capillary is raised and the process repeated at the header post and for the other pads on the chip.

One variation in this basic method eliminates heating and uses ultrasonic bonding instead. In this method the tungsten-carbide capillary is connected to an ultrasonic transducer. When in contact with a pad or post, the wire is vibrated under pressure to form the bond. Because of the ball at the end of the wire it is called a ball bond; a nailhead bond occurs when the shape of the ball is deformed after the bond is made.

Aluminum is used in ultrasonic bonding. It has several advantages. A major one is the absence of metallurgical reactions in bonds between gold wire and aluminum pads. When aluminum wire is used, the wire is bent under the edge of a wedge-shaped bonding tool. The tool then applies pressure and ultrasonic vibration to form the bond. The resulting flat bond is called a wedge bond.

Thermocompression and ultrasonic techniques are done manually under a microscope by an experienced operator; labor costs rise quickly in circuits with many terminals. A further disadvantage of wire bonding is that the heating and pressure involved can be harmful to some IC chips.

Flip-chip and *beam-lead* methods eliminate the disadvantages of bonding wires individually. In each case, relatively thick metal is deposited on the contact pads before the devices are separated from the wafer. After being separated, the deposited metal is used to contact a matching metallized pattern on the substrate.

In the flip-chip method, globules or bumps of solder or a special alloy are deposited on each contact pad. These metal bumps rise about

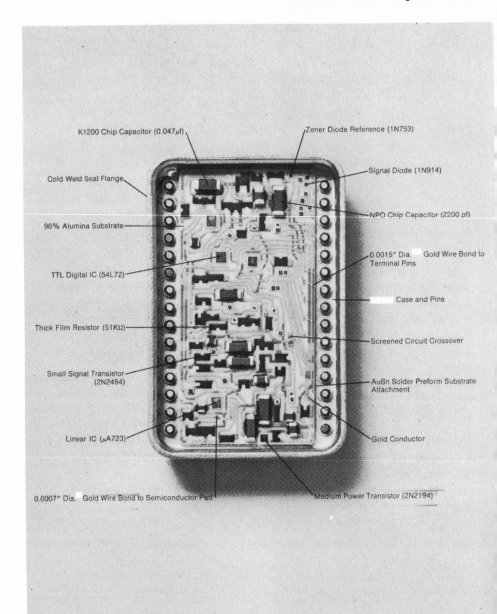

K1200 Chip Capacitor (0.047μf)

Zener Diode Reference (1N753)

Cold Weld Seal Flange

Signal Diode (1N914)

96% Alumina Substrate

NPO Chip Capacitor (2200 pf)

0.0015″ Dia. Gold Wire Bond to Terminal Pins

TTL Digital IC (54L72)

Case and Pins

Thick Film Resistor (51KΩ)

Screened Circuit Crossover

Small Signal Transistor (2N2484)

AuSn Solder Preform Substrate Attachment

Linear IC (μA723)

Gold Conductor

0.0007″ Dia. Gold Wire Bond to Semiconductor Pad

Medium Power Transistor (2N2194)

Photo courtesy of Beckman Instruments

Figure 7-9. A hybrid package can use a wide range of components—from linear to digital ICs and from thick film resistors to chip capacitors.

Figure 7-10. Chip carriers with multilayer ceramic substrates and dual in-line pack-ages illustrate the variety of constructions available to hybrid and other integrated-circuit device manufacturers. (Courtesy of the 3M Co.)

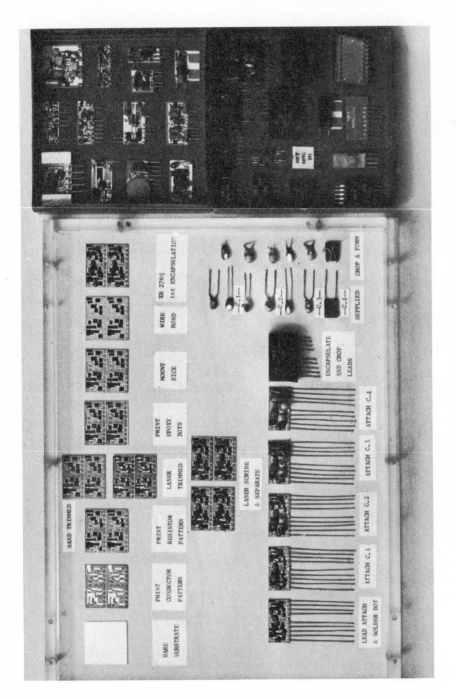

Figure 7-11. Even large, discrete components, such as capacitors, can be encapsulated within a hybrid package.

2 mils above the surface of the chip. After separation from the wafer, the chip is turned upside down and the bumps are properly aligned with the metallized pattern on the substrate; then ultrasonic bonding or solder alloying attachs each bump to its corresponding point on the substrate. All connections are made simultaneously. Because the bonds are made under the chip, however, they can't be inspected visually. Furthermore, it is still necessary to heat or exert pressure on the chip.

In beam-lead technology, however, bonds to the substrate pattern are made external to the chip. About ten-micron-thick metal tabs on the chip lead away from the circuit at each contact pad. These relatively large beam leads are commonly made by electroplating gold onto the wafer as defined by a photoresist pattern. Several etching steps then isolate the leads extending outside the active area of each circuit. Thus subsequent bonding does not apply heat and pressure directly onto the chip.

Although the beam-lead process requires many extra fabrication steps, protection of the chip from heat and pressure, the accessibility of the leads, the advantages of one-step bonding and excellent isolation between devices (see page 170), makes the beam-lead method worthwhile. Along certain crystallographic directions, air gaps as small as 0.2 mil can be etched between circuit elements. This procedure results in the high isolation between elements that is typical of large discrete components.

Packaging the chip

The final step in IC fabrication packages the circuit to protect it from the environment. When packaged in metal headers, the device is alloyed to the surface of the header, wire bonds are made to the header posts and a metal lid is welded over the assembly. Such complete sealing of the unit from the outside environment is called hermetic sealing. An inert gas, which maintains the device in a prescribed atmosphere, is often added to the package.

Plastic encapsulation is well suited for assembly-line processing and is less expensive than header mounting. Since plastic encapsulation provides mechanical support, flip-chip or beam-lead circuits need not be mounted on a substrate, but can be placed directly on a metal-lead pattern instead. Such a metal pattern can be stamped out of a continuous ribbon, with a metal edge to hold the pattern together. This edge is trimmed off after encapsulation to provide leads from the package.

Why separate the chips for interconnection on a separate substrate as in hybrid packages? They can be left instead, on the single wafer and interconnected by appropriate metallization. The result is the technology of large-scale integration (LSI).

An obvious problem is the complex pattern of required interconnections. Many crossings are required in the metallization pattern. Even with optimum interconnections, some crossings are inevitable. Multilayer metallization, with insulating layers between connection layers, can solve this problem.

An important difficulty is the presence of faulty circuits on the wafer. With a predetermined interconnection pattern, faulty circuits are automatically wired into the system; therefore, discretionary wiring must be used. Only good circuits are interconnected. Discretionary wiring requires more circuits on the wafer than are needed in the completed system. Each circuit is tested with a multiple-point probe and a special interconnection pattern selected for each wafer of circuits.

One method uses a computer to generate a pattern from which a photographic mask is prepared for the metallization. The acceptable circuits on the wafer are interconnected to perform the desired operation.

Defects, tolerances and yields

Defects in ICs can be separated into three categories: *area defects*, which cover a large portion of the slice; *spot and line defects*, which affect only a few elements or circuits; and *handling defects*, which occur after the circuit has been separated from the slice. Spot defects predominate: these include crystal dislocations, diffusion pipes and oxide pinholes. Area defects result from breakage, diffusion and masking of errors and contamination.

Along with defects, tolerances are also closely associated with yield—the percentage of good circuits produced on a slice.

ICs require a very large number of processing steps. In each of these steps, defects and tolerance errors can be introduced into the individual elements. In most circuits, one defective element means that the entire circuit is useless.

Even if manufacturing is carried out with utmost care and the percentage of good circuits after each step remains large, the number of circuits at the end of the line is surprisingly small. For example, if in each step only 2 percent of all circuits is lost, the yield for each step is 98 percent. After 10 processing steps the net yield is $Y = 0.98^{10} = 0.801$ and $Y = 0.98^{50} = 0.355$—35.5 percent of the total number of circuits will work after 50 steps. Yield is also greatly affected by the area of the chip. The probability of a defect increases exponentially as the area of the chip increases. Doubling the area increases the probability of the chip having a defect about eight times.

How large, then, should a circuit be? As many components as possible should be included in each circuit to minimize the number of individual circuits and the connections between them, but the larger the circuit, the lower the yield. The optimum number of elements per IC, however, increases as better process control is achieved. Thus no specific number can be given since progress in LSIs is rapid and far from having reached a saturation stage.

Chapter 8

Diodes Do Many Things
Other than Rectify

The most obvious property of a p-n junction is its directional conductivity. It conducts current best in only one direction. An *ideal diode* is a short circuit when forward biased and an open circuit when reverse biased (Figure 8-1(a)). The junction diode does not meet this description perfectly but in many circuits, junctions can be approximated by the ideal diode in series with other circuit elements to form an equivalent circuit. For example, most forward-biased diodes have an *offset* voltage which can be approximated in a circuit by a battery in series with the ideal diode (Figure 8-1(b)). The series battery keeps the ideal diode from conducting for applied voltages less than the offset voltage. The approximation to actual diode characteristics is improved by adding a series resistor to the circuit equivalent (Figure 8-1(c)).

Diodes rectify

Junction diodes for use as rectifiers should have characteristics as close as possible to that of the ideal diode. The reverse current should be negligible and the forward current should exhibit little voltage dependence. The reverse breakdown voltage should be large and the offset voltage in the forward direction should be small. Unfortunately, not all of these requirements can be met by a single device; compromises must be made in the design of the junction to provide the best diode for the intended application.

The unilateral nature of diodes is useful also for many other

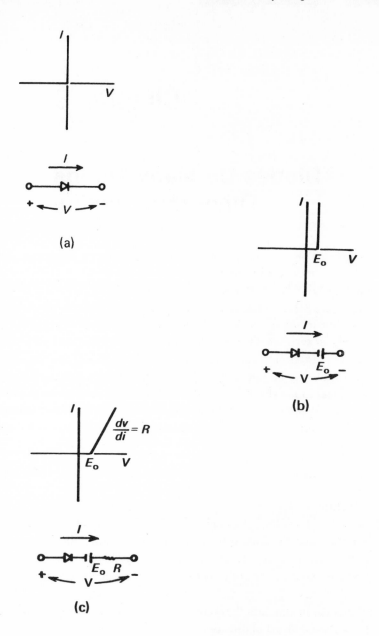

Figure 8-1. An ideal diode (a) has infinite reverse and zero forward resistance, but most practical junction diodes have an offset voltage E_o, which can be represented by a battery, as in (b), and some forward resistance, R, as in (c). Reverse resistance is usually ignored in most diode-equivalent representations.

circuit applications such as those that require *waveshaping* by the alteration of ac signals. The diode passes only certain portions of the signal while other portions are blocked.

Band gap[1] is obviously an important consideration in choosing a material for rectifier diodes. The reverse saturation current, which depends on thermally generated carriers, decreases with increasing bandgap. A rectifier made with a wide band gap material can be operated at higher temperatures.

On the other hand, the contact potential and offset voltage generally increase with band gap. For power work Si is generally preferred over Ge because of its wider band gap, lower leakage current and higher breakdown voltage, as well as its more convenient fabrication properties.

The *doping density* on each side of the junction influences the avalanche breakdown voltage[2], the contact potential and the series resistance of the diode. If the junction has one highly doped side and one lightly doped side (such as a p + −n junction), the lightly doped region tends to increase the forward resistance. To reduce the resistance of the lightly doped region it is necessary to make its area large and reduce its length. Physical geometry thus becomes another important design variable. The lightly doped region cannot be made arbitrarily short, however, because of an effect called *punch-through*. The result of punch-through is a reverse breakdown below the value normally expected.

In the fabrication of a p+-n or a p-n+ junction, it is common to terminate the lightly doped region with a heavily doped layer of the same type to make easier ohmic contact to the device. The result is a structure like p+-n-n+ with the p+-n layers serving as the active junction.

The lightly doped region determines the avalanche breakdown voltage. If this region is short compared with the minority carrier diffusion length, the excess carrier injection for large forward currents can increase the conductivity of the region significantly. This type of conductivity modulation reduces the forward resistance and is very useful for high-current devices.

---

[1]See Pages 19 to 23 and 119 to 121.
[2]See page 44.

Using the secondary properties of junctions

We have presented many of the properties of junction diodes in Chapters 3, 4 and 6. In this chapter we extend the treatment beyond the usual applications of diodes in rectification, switching and light sensitivity to include devices that depend on secondary junction properties.

So far we have discussed the properties of pn junctions made of relatively pure semiconductors. Impurity doping was only a small fraction of the total atomic density of the material. The few impurity atoms were so widely spaced throughout the sample that no charge transport could be said to take place within the dopants themselves. But if one continues to dope a semiconductor with impurities, a point is reached where impurities become packed so closely that interactions between them cannot be ignored.

For donors in high densities of 10^{20} donors/cm^3, the donor states form a band that overlaps the bottom of the conduction band so that the Fermi level, E_f, is no longer within the band gap, but lies within the conduction band. The material is then called *degenerate* n-type. Degenerate p-type material also occurs when the acceptor density is very high; then the Fermi level lies in the valence band.

In a degenerate n-type sample the region between E_C (conduction bands) and E_f is for the most part filled with electrons; in degenerate p-type the region between E_V and E_C is the forbidden energy gap, E_g.

If one of the diode's sides is made small and the depletion region across the junction is very narrow, then the electric field across the junction is large. The conditions for *electron tunneling*—filled and empty states that are separated by a narrow potential barrier of finite height—are thus established.

Tunneling is a quantum probability

Classical physics erroneously predicts that an electron can't penetrate a potential barrier unless the electron's energy is greater than that of the barrier; however, the theory of quantum mechanics shows that there is a small but finite probability that an electron, even with insufficient energy to climb the potential barrier, can penetrate a barrier that is sufficiently narrow. This phenomenon is called *tunneling*.

Figure 8-2(a) shows the energy diagram of a tunnel-diode junction

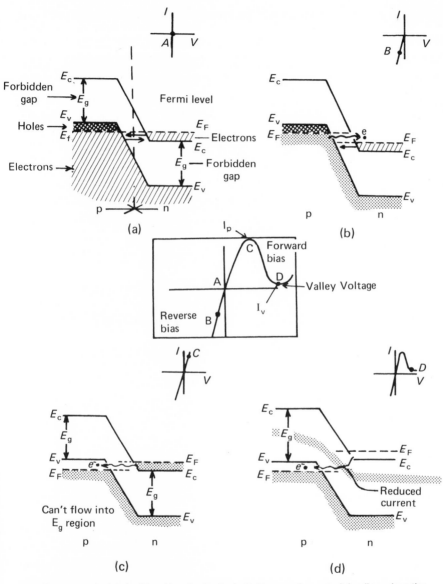

Figure 8-2. In (a) the Fermi levels, E_f, for both the p and n materials align when the tunnel is at zero bias. Reverse bias in (b) allows a heavy net flow of electrons from the valence bands of the p side to the empty conduction bands of the n side. For small forward bias, conduction and valence bands align so that only electrons from n to p side can flow (c). As forward bias is increased, band alignment is less, and less current flows from n to p sides (d). This produces negative-resistance region CD.

under zero-bias conditions. The cross-hatched regions represent those energy states in the conduction and valence bands which are occupied by electrons. The horizontal E_f line indicates the level to which the energy states on both sides of the junction are occupied by electrons. Conduction-band electrons from the n-type side tunnel into the p-type valence band and the valence-band electrons from the p-type side tunnel into the conduction band in equal and oppositely directed amounts so that net current is zero.

When a reverse bias is applied, the energy levels on the p side are increased in relation to those of the n side (Figure 8-2(b)). The p-type valence electron can easily flow from the p-type to the n-type side because the supply of electrons in the p valence band is so large and there is a large supply of empty states in the n conduction band that faces the electrons. The tunnel diode is thus highly conductive for all values of reverse bias. The current flowing from the n to the p side remains the same as in Figure 8-2(a).

For small forward-bias voltages, the energy levels of the p side are decreased in relation to the n-side (Figure 8-2(c)). The n conduction-band electrons are then opposite holes in the p side and the p valence-band electrons are now opposite the forbidden energy gap of the n side. The current from conduction to the valence bands remains the same as in Figure 8-2(c), but the current that flows in the opposite direction goes to zero.

As the forward voltage is gradually increased, more and more conduction-band electrons on the n side fall opposite the forbidden energy gap of the p side and the current is reduced. *Forward current thus decreases as the forward voltage increases.* This current-voltage relationship is called the *negative-resistance* region of the tunnel diode. For a forward-bias voltage equal to the *valley voltage*, tunneling completely ceases.

At applied voltages greater than the valley voltage, the height of the barrier is reduced to a level which permits conventional current to flow over the barrier. The characteristic then resembles that of a conventional diode. The forward current is now dominated by diffusion current—electrons surmounting the potential barrier from n to p and holes surmounting their potential barrier from p to n. Diffusion current is also present at low forward voltage in the forward tunneling mode, but it is negligible compared to the tunneling current.

Figure 8-3 compares the characteristic of a tunnel diode with that of a conventional rectifier diode. In the conventional diode, forward current does not begin to flow freely until a forward-bias voltage of about 0.3 to 3.0 V, sometimes referred to as the *offset* voltage, is applied. In the reverse-bias direction, the conventional diode has high resistance to current flow until breakdown is reached. The tunnel diode, on the other hand, is much more conductive near zero bias.

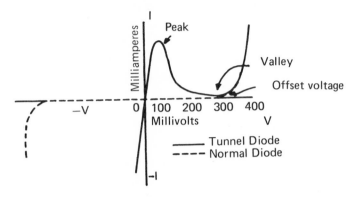

Figure 8-3. A normal diode has an offset voltage that must be overcome before forward current will flow. The forward current of a tunnel diode, however, starts with no offset voltage and its negative-slope region provides the diode's negative-resistance.

After the tunnel diode current reaches a sharp *peak*, it drops to a deep *valley*. This behavior occurs with a few tenths of volts of forward bias. The drop in current with increasing positive bias provides the negative-resistance characteristic of the tunnel diode. This property enables the tunnel diode to convert dc energy into ac energy and thus the diode can act as an amplifier or oscillator.

Important tunnel-diode specifications are the peak-point current I_p, the valley-point current I_v, the peak-to-valley current ratio I_p/I_v and the diode's capacitance and series resistance when biased in the negative-resistance region. Additional specifications are the maximum forward and reverse current ratings and typical values of the peak-point voltage, the valley-point voltage and the negative conductance. The peak-point current depends upon the junction area and ranges from a fraction of a milliampere to as high as several amperes. The peak-to-valley current ratio is a property of the semiconductor material. Typical values are of the order of 10:1 for germanium diodes and 4:1 for silicon diodes.

Tunnel diodes are used primarily in applications for which there are no suitable transistors. The tunnel diode is better than the transistor in several ways: it is basically a higher-frequency device because its electrical characteristics depend upon the propagation of electric waves across a space-change layer rather than upon the drift and diffusion of carriers. It is not dependent upon minority-carrier lifetimes and is therefore less sensitive to surface conditions and nuclear radiation. A tunnel diode is often not encapsulated and can be fabricated from materials that are unsuitable for other devices. It can be operated over a broader temperature range because of its heavier doping, and is basically less costly than the transistor.

Several disadvantages of the tunnel diode restrict its use. It is a two-terminal device that requires special circuitry. This makes it difficult to cascade amplifier stages. Its operating voltages are quite low—of the order of tenths of a volt. Although these voltages are large enough to switch transistors having base-emitter voltages of the same order of magnitude, they limit the tunnel diode to low-power applications. Another disadvantage is the problem of providing a suitable low-voltage power supply. Such a supply must have an extremely low internal resistance so that the diode can operate in the negative-resistance region and still have available appreciable power even at such a low voltage.

A tunnel diode's structure is simple

The structure of the tunnel diode is extremely simple. A small (approximately 2-mil) diameter, highly conductive dot of n or p type material is alloyed to a pellet of highly conductive oppositely doped material to form a junction. The pellet is soldered into a low-inductance, low-capacitance case. A fine mesh screen connects to the dot and the junction area is reduced by etching to produce the desired current density. The device is then encapsulated and a lid is welded over the cavity.

Tunnel diodes can be made from many semiconductor materials. They include germanium, silicon, gallium arsenide, indium phosphide, indium arsenide and indium antimonide. Materials with small forbidden energy gaps, low effective masses and low dielectric constants provide large tunneling probabilities. These materials permit the use of small junctions and low capacitance for given peak currents, and thus can provide extremely fast switching speeds.

Most commercial tunnel diodes are fabricated from either germanium or gallium arsenide. Germanium devices offer low noise and rise times as fast as 40 picoseconds (10^{-12} sec.). Gallium arsenide diodes have a larger voltage swing than germanium diodes. The other materials, in general, do not offer any advantages and usually have difficult fabrication or operation problems.

The high switching speed of tunnel diodes and their simplicity and stability make them suitable for computer applications. They can also operate effectively as amplifiers, oscillators and converters, especially at microwave frequencies. In addition, tunnel diodes have extremely low power consumption and are relatively unaffected by radiation and temperature.

The two-terminal configuration of a tunnel diode permits, on the one hand, the design of circuits that are extremely simple. On the other hand, the lack of isolation between input and output is usually a serious problem. At microwave frequencies, however, tunnel diodes have important advantages over transistors.

The load line determines the operation

The negative resistance of a tunnel diode can be used in many ways to achieve oscillation, amplification, switching and other functions. A simple bias circuit can produce several types of load lines (Figure 8-4). When the slope of the load line is less than the slope of the negative resistance region, three points intersect with the diode's characteristic. Points a and c are stable, but the circuit can't maintain the condition represented by point b. Any small increase in I causes V to decrease. This causes a further increase in I until the stable point a is reached. Similarly, a small decrease in current at b causes an eventual shift to the stable point c. Because of the two stable operating points, the circuit is called *bistable*.

If the circuit is in state a, a positive pulse that can overcome the peak current of the characteristic shifts the operating point to c. A negative pulse shifts it back to a. Applications such as high-speed memory-storage elements are thus possible.

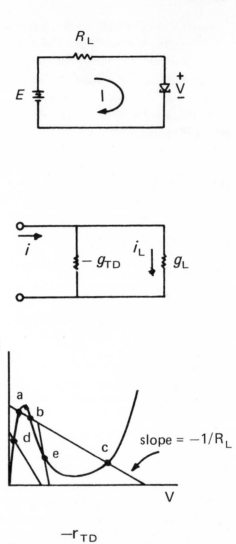

Figure 8-4. A simple load resistor and dc voltage source can bias a tunnel diode for several modes of operation.

If the load line crossing is in a positive resistance region d, the circuit is called monostable. A crossing in the negative resistance region e, is used in astable circuits. The astable line is used for amplifiers and oscillators, especially at microwave frequencies.

Microwave two-terminal devices operate on negative resistance

The tunnel diode and all other active two-terminal devices to be discussed later are negative-resistance devices. In a positive resistance, the current through the resistance and the voltage across it are in phase. A current that flows through a positive resistance produces an in-phase voltage *drop* and power is dissipated by heating the resistance. In a negative resistance, however, the current and voltage are 180 degrees out of phase and a current drop produces a voltage *rise*, which adds to the original voltage source to even further reduce the current, or vice versa. Thus positive resistances (passive devices) dissipate power, and negative resistances (active devices) can convert dc into ac power when connected into resonant circuits, by causing oscillations. A small voltage or current change is increased until something limits it, and the resonant circuit then reverses the direction of the change.

In microwave applications, the active device is usually mounted at the end of a transmission line (Figure 8-5). If there is a net negative

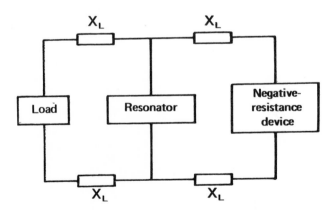

Figure 8-5. A two-terminal negative-resistance device is usually mounted at the end of a transmission line.

resistance in such a circuit, the circuit will break into oscillation at its resonant frequency. With the negative resistance kept below the oscillation point, such a circuit acts as an amplifier.

Tunnel diodes make good low-voltage backward rectifiers

Because of their offset voltage, conventional semiconductor diodes, particularly silicon diodes, do not work well near the origin of the conduction characteristic. Silicon diodes at room temperature act as virtual open circuits to voltages of either polarity unless the magnitude exceeds several hundred millivolts (Figure 8-6). A version of the tunnel diode can operate a rectifier near the origin.

If the doping places the n-region Fermi level at the bottom edge of the conduction band and the p-region Fermi level is at the top edge of the valence band, the bands overlap with the smallest reverse bias. No overlapping occurs with either zero bias or forward bias, however. Such a device, with a barrier thin enough for tunneling, has the reverse conduction characteristic of the tunnel diode, but a normal forward characteristic and the absence of tunneling.

The diode thus conducts extremely well for one polarity of applied voltage and quite poorly for voltages of the opposite polarity, but only for voltages less than several tenths of a volt. Since it operates with its normal anode and cathode regions interchanged, the device is known as a *backward diode*. The forward and reverse conduction characteristics of a backward diode are shown in Figure 8-6(a) and compared with a conventional rectifier.

Since some tunneling does occur when tunnel pn junctions are forward biased, the reverse characteristics of backward diodes are poorer than those of conventional diodes, but the backward-diode forward characteristics are much better. This is particularly evident in the silicon-diode comparison of Figure 8-6(b); however, the reverse voltage of the backward diode is limited to a value of about 1 V.

Zeners are also heavily doped diodes

In Chapter 3 reverse-bias breakdown voltage was discussed briefly. The region beyond this voltage was referred to as the Zener region. The reverse-bias breakdown voltage of a junction can be varied

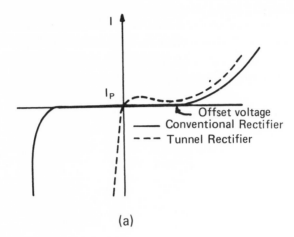

(a)

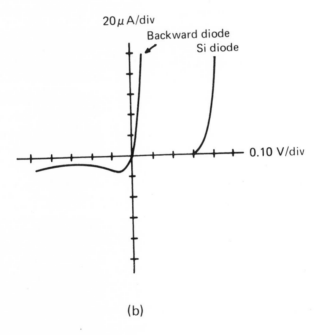

(b)

Figure 8-6. Special tunnel diodes, when operated in the reverse direction as compared to conventional diodes, can provide rectification at very low voltages. Conventional diodes have an offset voltage that must be overcome before conductions in the forward direction can start.

by choice of junction doping densities. The breakdown mechanism for abrupt junctions which occurs with extremely heavy doping is called the Zener effect, which is a tunneling effect. The more common breakdown is called avalanche or impact ionization in lightly doped junctions. Doping density can vary the breakdown voltages from less than one volt to several hundred volts. In well-designed junctions, the breakdown is sharp and the current after breakdown is essentially independent of voltage (Figure 8-7). Such diodes are usually called *Zener diodes* for both light and heavy doping even though they have different breakdown mechanisms.

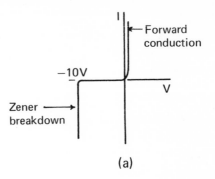

(a)

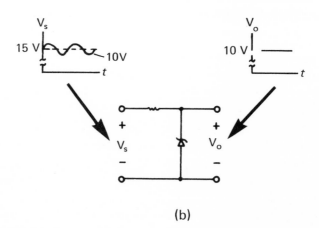

(b)

Figure 8-7. When operated in the Zener-breakdown region, a diode maintains an almost constant voltage drop across itself (a). This effect finds applications in simple power-supply voltage-regulator circuits (b).

Zener diodes are used as voltage regulators in circuits with varying inputs. The diode of Figure 8-7 holds the circuit output voltage constant at 10 V, while the input varies at voltages greater than 10 V. A rectified and partially filtered signal with a 15-V dc component and ± 1-V ripple remains constant at 10 V. More sophisticated regulator circuits use Zener diodes as *reference diodes*.

Tunnel and Zener diodes are related

There is a continuous variation of diode properties with doping —from the conventional diode to the tunnel diode (Figure 8-8). For simplicity, Figure 8-8(a) through (c) shows a constant degenerate doping density in the p region and the density of donors varies only on the n side of the junction.

As the doping density is increased, the Fermi level on the n side approaches closer to and then passes the bottom of the conduction band so that a smaller reverse voltage can cross the bands and provide a control of the Zener breakdown voltage (Figure 8-8(b)). In Figure 8-8(c), still more n doping brings the Fermi level very close to the edge of the n-conduction band and a backward diode characteristic is produced. When the doping causes the Fermi level to cross into the conduction region (Figure 8-8(d)), tunneling characteristics are achieved.

Internally generated negative conductances

Tunnel diodes exhibit a clear negative-resistance region in their current voltage plot; however, in some diode devices, the negative-resistance effect occurs when the device's internal local current density is properly out of phase with the local internal electric field. In the impact avalanche transit time (IMPATT) diode (Figure 8-9), often called a Read diode, the negative conductance is derived from a combination of avalanche multiplication and transit time effects. Diodes are reverse biased to avalanche breakdown when an ac voltage is superimposed on the dc bias. The reverse electric field in the junction generates many electron-hole pairs. Since the rate of ionization at a given instant depends not only on the magnitude of the electric field but on the number of charge carriers present, a current pulse, once started, continues to build up even after the voltage has begun to drop because of the exponentially increasing (avalanching) number of charge carriers. Current increase can continue as long as the applied

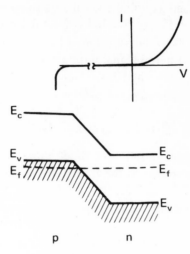

a) Conventional diode

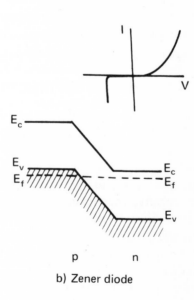

b) Zener diode

Figure 8-8. Gradually increased doping can change a diode's characteristics from conventional (a) to tunnel-diode performance (d).

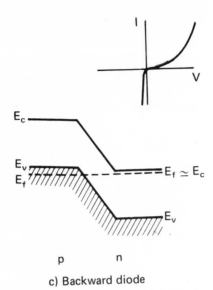

c) Backward diode

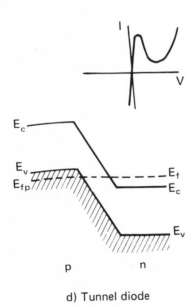

d) Tunnel diode

Figure 8-8 (cont.). For simplicity, in Fermi diagrams (a) to (d), the doping on the p side is kept constant and only the n side is increased.

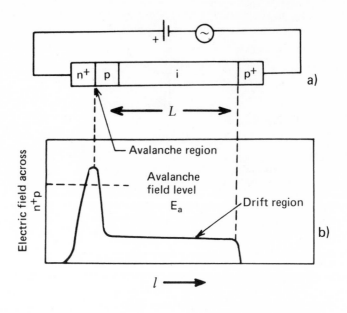

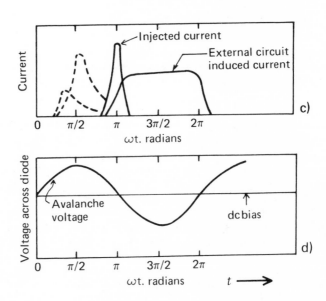

Figure 8-9. A Read IMPATT diode achieves its negative resistance internally by a combination of avalanche multiplication and transit-time delay.

voltage remains above the avalanche level, even though the voltage is decreasing.

The carriers generated by the avalanche process are swept through a drift region to the terminals of the device. The ac component of the diode current can be approximately 180 degrees out of phase with the applied voltage, producing negative conductance and oscillation in a resonant circuit. IMPATT devices can thus convert dc to microwave ac signals and with high efficiency.

Assume that the p region is very narrow and that all the avalanche multiplication takes place in a thin region near the n+ −p junction. If the dc bias and superimposed ac is such that the critical field for avalanche is met in the n+ −p space-charge region, avalanche multiplication begins. Electrons move to the n+ region and holes enter the i-drift region.

With the device mounted in a resonant microwave circuit, an ac signal can be maintained at a given frequency. As the voltage goes positive the multiplication process continues to grow as long as the electric field is above E_a. The particle current from the avalanche increases exponentially with time while the field is above the critical value. The important result of this growth is that the hole pulse reaches its peak value not at $\pi/2$ when the ac voltage is maximum, but at π (Figure 8-9(c) and (d)).

There is a phase delay of $\pi/2$ because of the avalanche process itself. The remaining delay is provided by the drift region. When the avalanche multiplication stops, the pulse of holes drifts toward the p+ contact. During this period the ac voltage is negative and the ac plus dc is at a minimum; thus minimum voltage and maximum current mean that the dynamic conductance is negative and energy is supplied to the ac field.

If the length of the drift region, L, is chosen properly, the pulse of holes is collected at the p+ contact just as the voltage cycle is completed, and the cycle then repeats itself. The optimum frequency, therefore, is one-half the inverse transit time of holes across the drift region.

The parameter L is critical. For a drift velocity 10^7 cm/sec for Si, the optimum operating frequency for a device with an i region length of 5 μm is f = $10^7/2(5 \times 10^{-4})$ = 10^{10} Hz. Negative resistance is exhibited by an IMPATT diode for frequencies somewhat above and below this optimum frequency for exact 180° phase delay.

Other IMPATT structures are simpler

Although the Read diode of Figure 8-9 is most direct in explaining the operation of IMPATT devices, simpler structures can be used. In some cases they may be more efficient. Negative conductance can be obtained in simple $p-n$ or $p-i-n$ devices. In the case of the $p-i-n$, most of the applied voltage occurs across the i region, which serves as both a uniform avalanche region and a drift region. The two processes of delay and avalanche—separate in the Read diode—are here distributed within the i region of the $p-i-n$; thus, both electrons and holes participate in the avalanche and drift processes. This type of operation is called a TRAPATT cycle, for trapped-plasma avalanche-triggered transit.

A large overvoltage causes an *avalanche zone* to move through the i region and fill the region with an electrostatically neutral electron-hole population or *plasma*. The plasma is created very rapidly, but the carriers drift slowly. Initially this large numbers of carriers greatly reduces the terminal voltage across the diode. As holes and electrons drift apart, the neutral plasma is depleted, the terminal voltage builds to a high value and the current goes toward a low value. An increasing voltage that produces a reduced current, and vice versa, is a negative conductance. In a proper resonant circuit there is a cyclical buildup and discharge of the plasma that gives rise to efficient microwave power generation.

BARITT diodes—barrier injected transit-time diodes—have long drift regions similar to those of IMPATT diodes; however, the carriers that traverse the drift region of a BARITT diode are minority carrier injections from a forward-biased junction rather than from the plasma of an avalanche region. BARITT diodes have several different types of structures. They include p-n-p, p-n-metal and metal-n-metal. They are much less noisy than IMPATT diodes and can operate with much lower voltages. The present major disadvantages of BARITT diodes are relatively narrow bandwidth and power outputs limited to a few milliwatts.

Carrier drift velocity can have a negative slope

In most bulk semiconductors current carriers reach a scattering-limited velocity, and the velocity-vs-field plot saturates at high fields

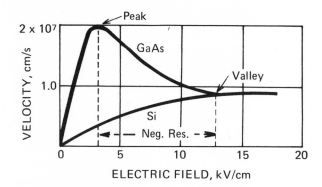

Figure 8-10. The drop in drift velocity of GaAs with increase in electric field constitutes a region of negative resistance. This phenomenon results from an apparent increase in the effective mass of carrier electrons as they transfer from lower to higher conduction energy bands. Gunn diodes use this effect in microwave oscillators. Note the comparison with the drift-velocity curve of Si, which does not exhibit the Gunn effect of GaAs.

as for Si (Figure 8-10). In some materials, however, the energy of electrons can be raised by an applied field from one region of the conduction band to another region with higher energy. For some band structures, negative conductivity can result from this electron transfer.

Microwave devices which operate by the *transferred electron* mechanism are often called *Gunn diodes* after J.B. Gunn who first demonstrated one of their forms of oscillation. There are many modes of operation for these devices. In the transferred electron mechanism, the conduction electrons of some semiconductors are shifted from a state of high mobility to a state of low mobility by the influence of a strong electric field.

This transfer to low mobility produces a negative slope in a part of the drift velocity-electric field curve of GaAs (Figure 8-10). If a GaAs Gunn device is biased in this region of the curve, an increase in voltage will cause a decrease in the drift velocity and consequently in the current. This effect is again equivalent to a negative resistance which can initiate microwave oscillators.

A large differential of velocities in the negative velocity-electric field region between the peak velocity and the minimum or valley velocity is highly desirable. A large "peak-to-valley" ratio means that a large current swing or negative resistance is obtainable.

Index

J

L

M